Edan Lepucki is a graduate of the Iowa Writers' Workshop and is a staff writer for *The Millions*. Her short fiction has been published in *McSweeney's* and *Narrative* magazine, among other publications, and she is the founder and director of Writing Workshops Los Angeles. This is her first novel.

'[A] subtle, melancholy picture of an America that's crumbling into oblivion. In *California*, the apocalypse is ongoing and worryingly plausible' *SFX*

'Breathtakingly original, fearless and inventive, pitch perfect in its portrayal of the intimacies and tiny betrayals of marriage, so utterly gripping it demands to be read in one sitting: Edan Lepucki's *California* is the novel you have been waiting for, the novel that perfectly captures the hopes and anxieties of contemporary America. This is a novel that resonates on every level, a novel that stays with you for a lifetime. Read it now' Joanna Rakoff, author of *My Salinger Year*

'An expansive, full-bodied and masterful narrative of humans caught in the most extreme situations, with all of our virtues and failings on full display: courage, cowardice, trust, betrayal, honor and expedience. The final eighty pages of this book gripped me as much as any fictional denouement I've encountered in recent years ... I firmly believe that Edan Lepucki is on the cusp of a long, strong career in American letters' Ben Fountain, author of *Billy Lynn's Long Halftime Walk*

'Thrilling and thoughtful ... a vivid, believable picture of a not-so-distant future and the timeless negotiation of young marriage, handled with suspense and psychological acuity' Janet Fitch, author of *Paint it Black*

'It's tempting to call this novel post-apocalyptic, but really, it's about an apocalypse in progress, an apocalypse that might already be happening, one that doesn't so much break life into before and after as unravel it bit by bit. Edan Lepucki tells her tale with preternatural clarity and total believability' Robin Sloan, author of *Mr. Penumbra's 24-Hour Bookstore*

'Stunning ... a book as terse and terrifying as the best of Shirley Jackson, on the one hand, and as clear-eyed and profound a portrait of a marriage as Evan Connell's *Mrs. Bridge*, on the other. *California* is superb' Matthew Specktor, author of *American Dream Machine*

'[A] stunning and brilliant novel, which is a wholly original take on the post-apocalypse genre ... funny and heartbreaking, scary and tender, beautifully written and compulsively page-turning, this is a book that will haunt me, and that I'll be thankful to return to in the years to come. It left me speechless. Read it, and prepare yourself' Dan Chaon, author of *Await Your Reply*

'*California* is a wonder: a big, gripping and inventive story built on quiet, precise human moments. Edan Lepucki's eerie near future is vividly and persuasively imagined. She is a fierce new presence in American fiction' Dana Spiotta, author of *Stone Arabia*

'Lepucki's story is a compelling examination of personal relationships laid bare. A provoking thought experiment, her novel imagines a chillingly plausible future' *Huffington Post*

'Lepucki has armed her novel with a stunning twist, and its fallout is thrilling' *Washington Post*

'Lepucki focuses on the complexities of basic human emotions, testing allegiances and letting secrets unravel even the most steadfast of survivors, all while illustrating how impossible it is to change what inherently makes us human' *Los Angeles Times*

CALIFORNIA

EDAN LEPUCKI

ABACUS

ABACUS

First published in Great Britain in 2014 by Little, Brown
This paperback edition published in 2015 by Abacus

1 3 5 7 9 10 8 6 4 2

A CIP catalogue record for this book
is available from the British Library.

ISBN 978-0-349-13947-0

Typeset in Bembo by M Rules
Printed and bound in Great Britain by
Clays Ltd, St Ives plc

Papers used by Abacus are from well-managed forests
and other responsible sources.

MIX
Paper from
responsible sources
FSC
www.fsc.org   FSC® C104740

Abacus
An imprint of
Little, Brown Book Group
100 Victoria Embankment
London EC4Y 0DY

An Hachette UK Company
www.hachette.co.uk

www.littlebrown.co.uk

For Patrick

I

On the map, their destination had been a stretch of green, as if they would be living on a golf course. No freeways nearby, or any roads, really: those had been left to rot years before. Frida had given this place a secret name, the afterlife, and on their journey, when they were forced to hide in abandoned rest stops, or when they'd filled the car with the last of their gasoline, this place had beckoned. In her mind it was a township, and Cal was the mayor. She was the mayor's wife.

Of course it was nothing like that. The forest had not been expecting them. If anything, it had tried to throw them out, again and again. But they had stayed, perhaps even prospered. Now Frida could only laugh at the memory of herself, over two years ago: dragging a duffel bag behind her with a groan, her nails bitten to shit, her stomach roiling. Grime like she'd never imagined. Even her knees had smelled.

She thought it would be easier once they arrived; she should have known better. The work didn't end then; if anything, it got worse, and for months the exhaustion and fear tick-ticked in her body like a dealer shuffling cards. At night, the darkness gave her a skinned-alive feeling, and she longed for her old childhood bed. For a bed, period.

She had packed some things to comfort herself: the dead Device, a matchbook from their favorite bar. Cal later called them her *artifacts.* In a world so disconnected from the past, her attachment to these objects had been her only strategy for remaining sane. It still was.

She tried not to take them out too often, but Cal had left the house to do some digging, and he wouldn't be back for at least an hour. Even though the sky was gray, the sun weak, he'd worn his plaid button-down and a bandanna around his neck. They still had a bottle of sunscreen, but it had expired and was as watery as skim milk.

"Stay inside for a while," he'd said before he left.

Frida had linked her arms around his neck. "Where would I go?"

He kissed her goodbye on the mouth, as he still did, and always would. She was thinking, already, of the artifacts tucked away in an old briefcase, shoved under one of the unused twin cots. It had been a rough morning.

Cal had latched the door behind him, and once his footsteps receded, she went right for the briefcase. From the small pile of artifacts, she picked up the abacus. She liked to pull the blue beads back and forth across the wires. She counted, she tapped, she closed her eyes. Frida had played with the abacus as a little girl and even then had depended on its calming effect. Her brother Micah, two years younger, had one as well, red beads instead of blue, but one day, when he was about seven, he cut it apart and strung the beads onto a piece of yarn. He'd presented it to their mother as a necklace.

Frida flicked at the beads. She found herself counting the days yet again. Forty-two.

"I'm late," she said aloud, and her voice in the one-room house sounded small and plaintive. The walls seemed to breathe in the words; they would keep the secret until she told Cal.

"I'm late," she repeated, and willed her voice to stay steady.

2

She'd have to tell him soon, and like this. She could not be freaked out. She would have to declare it, as she would any fact.

Frida pulled the last bead across the abacus. It would be pleasurable, she thought, to pluck the wire from the frame and let the beads fall. She would pop one in her mouth and suck on it like a candy. But then she wouldn't have the abacus.

She put the thing down and sifted through the briefcase for something better. The other artifacts wouldn't do. Not the Device, nor the matchbook, nor the ripped shower cap she couldn't stand to part with. Not her mother's handwritten cake recipes, already memorized and useless out here. Not the box of antique pencils, nor her bottle of perfume, halfway empty.

She knew what she wanted.

Unlike the other objects, the turkey baster had been new. She'd brought it with them precisely because they hadn't had one in L.A.; it was something different, a simple object to mark a before and an after. She had liked the idea of using it at Thanksgiving, although she hadn't been sure they'd celebrate that anymore. She didn't think there would be turkeys here, and she'd been right.

Thanksgiving. That holiday was so quaint in her memory it felt like something from a storybook: *Once upon a time, Goldilocks ate herself silly.*

Frida couldn't hold herself back any longer and pulled the baster out of the briefcase. It was stored in old Christmas wrapping paper, printed with gingerbread men and mistletoe, and she unwrapped it slowly. She had last looked at the baster a few weeks ago, and she had taken care to put it back properly. It could not be damaged.

At the store, Frida had so much fun playing with the turkey baster, squeezing its plastic bulb so that the air farted out the glass tip. Frida had wondered if they might use it to try to get pregnant someday: their own ad hoc fertility treatment. It was funny how that had been on her mind even then.

But, no, Frida thought now, she wasn't pregnant. Couldn't be. She'd stop thinking about it.

The baster had been on sale. The store, like so many others, was going out of business. When the first of them perished, it had seemed impossible. "A chain like that!" people had said. When she was younger, Frida used to go there with her friends to marvel at all the useless necessities: the soy sauce receptacles, the tiny mother-of-pearl spoons, the glass pitchers. Even then, she didn't know anyone who could afford such things. When she turned thirteen, she spent all of her birthday money on a single cloth napkin. Her mother would have killed her had she known; things weren't dire then, not yet, but times were tough, and Frida could imagine her mother decrying such waste. Frida had stored the napkin in the pocket of a coat she never wore.

But on her last visit, at twenty-six, she was no longer that same stupid little girl, or so she told herself. The place had been ransacked. Frida still remembered the starkness of the flood-lights; they ran on a generator in the corner, illuminating the remaining coves of products, which were jumbled together in plastic bins. The register was by the entrance, and the girl who worked there accepted gold only, and not jewelry—it had to be melted down already.

Frida couldn't conjure the girl's face anymore, but she did remember her eyeliner. How had she gotten her hands on eye-liner? Perhaps it was an old stick of her mother's, gone to crayon at the back of the medicine cabinet. She could have sold it, if she wanted to, but she hadn't. The girl was barely eighteen, more likely sixteen. The place shut down a week later, didn't even make it to Christmas.

By the following spring, Frida was celebrating her twenty-seventh birthday in an empty apartment, their belongings packed and ready by the door. She'd wanted to spend one more in L.A.; she'd been born there, after all. Cal couldn't argue with that.

Frida held the baster by its plastic bulb, lifting it above her head. She imagined the store had probably gone feral soon after they left, like the rest of the businesses at that stupid outdoor mall. The Grove, it was called. Maybe in these two years it had sprouted some trees, finally earned its name. The famous trolley, rusted, its bell looted. The fountain, which had once lured tourists and toddlers to its edge, was probably dry; that, or sludgy with poison.

But what about the girl? Maybe she had been brave and stupid enough to head for the wilderness with only a bag full of tiny sherry glasses and cloth napkins to keep her company. Maybe a turkey baster, too.

Back in L.A., Frida had kept the baster a secret from Cal because she'd spent gold on it, gold they were saving for their journey. They'd saved for almost a year to get enough money for gas and other supplies. She had purchased something frivolous, and she knew it. She was still that same little girl, hoarding her treasure. She hadn't changed at all.

Once they were leaving, she kept the baster a secret because she was afraid Cal would say they couldn't bring it with them. They could only fit so much in the car, and before it ran out of gas, they would have to abandon it, carry their possessions the rest of the way. There was so much to carry, they had ended up making multiple trips with their stuff, and then they drove the car in the opposite direction until it sputtered dead, so they couldn't be followed. It was a small miracle that they found their possessions again, piled where they'd left them, unharmed.

Frida had smuggled the baster, like she had most of the artifacts. Cal eventually discovered her other things, but she'd still managed to keep the baster hidden.

She'd initially intended on using it in the afterlife, in whatever way it was most needed. And then, one day, she realized she wouldn't. Occasionally she toyed with the idea of snapping off the tag, which was attached to a string at the base of the bulb. At least it wasn't one of those plastic threads; she used to hate those, how they would leave holes in clothing and require scissors to remove. Those doodads were probably the whole reason America had gone to hell, the plastic seeping poisons, filling up landfills. What foolishness. But she loved the turkey baster precisely because it still had its tag. She loved its newness: the pure glass of the cylinder, its fragility, and the plastic butter-yellow bulb still chalky to the touch. It inhaled and exhaled air like that first time. She had to keep it hidden. It belonged only to her, and the secret of it had become as precious as the object itself.

Frida was tucking the briefcase under the bed when Cal stepped back into the house, ducking to get through the oddly small door. She liked how tall her husband was, and his narrow shoulders made him look even taller: stretched. Every morning he combed his short reddish hair with his fingers; it was so fine that little knots formed at the back of his head as he slept, and he hated it. Frida loved that, and she loved how every morning he woke with crescent-moon bags under his golden-hazel eyes, no matter how well rested he was.

A fine veil of soil covered his shirt and face, and he'd untied the bandanna from his neck so that he could wipe the sweat from his brow. The room filled with the sweet stink of him. Their feet had started to smell—not the vinegary scent that had cursed Frida in L.A., but something fungal and rotting, a bag of dying vegetables. Cal had said they smelled homeless, and she agreed. That's when they brought out their last Dove bar and their tube of antifungal cream. They didn't discuss what would happen when they ran out.

Their homemade soap, made from Douglas fir and the fat of vermin, smelled great but didn't actually work.

"How are the traps?" Frida asked.

Cal shrugged and went toward the thermos. They drank coffee once every two months, a treat, and the rest of the time they filled the thermos with water from the well. On the morning after a coffee day, the water absorbed some of the bitterness that still coated the thermos. If the world didn't end, and they moved back home, she would sell it to the cafés, get rich off coffee-water.

Cal filled his cup and drank it in one gulp, his Adam's apple sliding up and down his neck. That Adam's apple. He had once explained to Frida how Plato believed that the soul's parts—its reason, its passion—were located all over the human body. Frida liked to imagine Cal's soul, a sliver of it, residing in his slender neck, the jagged cliff that signified he was a man. He could never pull off drag with an Adam's apple like that.

"I know you think the traps are ridiculous," he said when he was finished drinking.

"I don't. You've built dozens of snares before, and they've worked. Why would I question you on traps?"

"You didn't have to."

"It's not like I rolled my eyes," she said, approaching him. Cal did stink. She handed him a towel from the shelf of supplies and told him to wipe off.

He gestured to the holes out the open door. "These traps will be bigger than usual, I know. But those gophers are stupid. They're bound to run in there."

"But, babe, this isn't *Robinson Crusoe*. Do you even know how to build a trap?"

He removed his shirt, so he could clean off his pits. "I did it as a kid," he said.

Frida sighed. "For fun or for real?"

"What's the difference?" he asked.

"You're lucky you're so clever," she said, and kissed him on the cheek.

He'd been working so hard out there. Maybe the holes he was digging would also keep them safe from the bigger beasts: the coyotes, the bears, and the wolves, which they sometimes heard howling at night.

Cal had been designing the traps in his journal for a few days, the physics and all that. He said they had worked on his father's land when he was a kid, and they would work now. Frida wasn't sure what gopher meat would taste like, but Cal said, "Protein is protein," and they couldn't be picky. They'd eaten a snake once—Bo Miller had cooked it for them—and occasionally they craved that, especially in winter, after days of turnips and potatoes.

Frida took the towel from him. "I know you're dying for meat," she said. "I've heard gophers taste like steak."

He sighed. "If I could just stop *wanting* it . . ."

"If only," she said.

He was already on his way out again. Back to work.

"Wait," she said. Would she tell him now? Forty-two days, she thought.

"What is it?" he asked.

"August's supposed to come this week. Should we see if he's got soap?"

"He never has soap."

That was true. For over a year, August had been a fixture in the afterlife, something to mark the time by. He arrived once a month on his mule-drawn buggy with goods to trade and information to gather. He wanted to know how they were feeling, and he liked to share notes about the weather, too. Once Frida had a cold, and he'd asked her what color her snot was.

"Clear," she'd told him.

He'd smiled and said, "It's going to be real cold tonight, so bundle up."

Frida had once traded August an acorn squash for a dented tin of evaporated milk and, another time, her old cashmere sweater for a knife, recently sharpened. As he handed it to her, blade down, he'd said, "For cooking, or weaponry." A statement, not a question, for it was understood that all tools in the wilderness needed to be versatile.

August was a thin black guy, probably ten years older than they were, just shy of forty, and he wore the never-quite-faded desperation of a former addict. "A tendency toward the vampiric" was how Cal had once put it. August even called himself a junkie, and he was: he traded junk for other junk. He liked to say he was the last black man on earth, and he might have been; around here, all jokes looped back to sour.

"I want to try planting some garlic," Cal said. "Maybe he has some."

"Okay."

"There's that look again. What is it?"

"It's nothing. Go digging."

"Whatever it is you're worrying about, just don't."

She said she'd try not to.

Cal waved at her from the doorway.

"Breathe!" he called out behind him.

Frida exhaled. How could he tell?

He'd been saying that for as long as she could remember.

He'd said it a lot during those first few months out here. He had kept her calm. Occasionally, his own nervousness about their survival spiked, and the air around him tightened, but most of the time, he seemed almost peaceful. It was as if he'd just returned from a monastery, his eyes gentle and open to the world, its good and its evil, the fair and unfair. Meanwhile, she could not even remember to breathe. It had taken everything to keep herself from saying,

We'll die out here, won't we?

Back then, she and Cal were living in the shed, and they thought they might be there for good. Neither knew that they'd eventually have a house to move into.

They'd stumbled upon the shed, searching for a good spot to settle, and its presence had saved them. The truth was, they had been clueless, some might even say reckless, about their plan. They were headed for open space, and that was all. "I just want to go away," Cal had first said to her. "I can't stand how awful everything is here."

Because she understood, Frida hadn't asked him to elaborate. He could have meant L.A.'s chewed-up streets or its shuttered stores and its sagging houses. All those dead lawns. Or maybe he meant the closed movie theaters and restaurants, and the parks growing wild in their abandonment. Or its people starving on the sidewalks, covered in piss and crying out. Or its crime; the murder rate increased every year, and the petty theft was as ubiquitous as the annoying gargle of leaf blowers had once been. The city wasn't just sick, it was dying, and Cal had been right, it was awful.

The shed had been a sound-enough structure: the walls, floor, and ceiling made of wooden planks, a roof covered by six tires, held together with baling wire. Cal had said, "Let's move in," to which Frida had replied, "Yeah, sure, nice outhouse." But she knew this shed was better than anything the two of them would be able to build on their own. Cal had done construction on his father's farm and, a little later on, in college, but he'd never built a home.

"I can do it," he'd told her as they moved their stuff into the shed. He said they could sleep there as they built an expansion. "I can do it with your help."

"That's what I'm worried about," Frida answered. "You and me, alone."

At first, that's how it had been. August hadn't found them yet, nor had the Millers, their closest and only neighbors, a few miles to the east. They later learned that Bo Miller had built the shed, years before. Their first four months out here, Cal and Frida had spoken only to each other, and sometimes that was the hardest thing, more trying than the planting or irrigating or the labor it took to build the rudimentary outdoor kitchen. Though she'd tried to prepare herself, Frida couldn't believe that they were really alone. Just the two of them.

One afternoon, at the end of their first summer, Cal had just called her over to the shower, a plastic receptacle heated by the sun that they'd secured to a tree branch. They had done this back home, when the gas bills got too high, although they'd hung the warmed water in the shower stall. Now they were outside. Everything was outside; it was like they were on an eternal camping trip.

That day the air was still warm, but with a sharpness to it that hinted at the chill to come. Frida looked forward to autumn; she actually liked collecting wood and making a fire as Cal had taught her to do. It seemed almost romantic. But Cal had warned her that she didn't really know what cold felt like. And he was right; she didn't.

"Go ahead," Cal had said, his hand on the plastic. He was

confirming its temperature, and all she had to do was turn the plastic spigot.

Frida thanked him and pulled her dress over her head. She no longer bothered with underwear or a bra. She liked being naked outside. Right then she tried to catch her husband's eyes, maybe shimmy her shoulders and bite her lower lip. Remind him how nice the line of her hips was. She might even say, *Hey there,* and smile.

But Cal had already turned away. He had the next task on his mind—the first one, perhaps, being his wife. In their four months out here, Frida had become a problem to solve, and once solved, she was invisible to him.

At the time, Frida imagined herself describing the moment. Maybe to an old friend or to her mother. Or online, as she used to do until their last year in L.A., before electricity became too expensive, before the Internet became a privilege for the very few. She had once kept a diligent online record of her life; she'd had a blog since she'd been able to write. Her brain couldn't just let that habit go, and in her head she said, *There I was, naked, my hair falling over my shoulders. But he didn't care! He had become immune to my nakedness.* The phrase was so silly, so melodramatic. *Immune to my nakedness.* But it was true. Cal wasn't looking.

And all at once she understood: no one was looking.

That day, Frida stood under the weak stream of water, never as hot as she wanted. It was the end of summer, and the only thing this world could promise them was that it would get colder, which would certainly crush their morale further. The finality of their situation sat on her chest like a brick and pushed. No one was looking. Her audience was sucked away, the ones keeping her safe with their concern, keeping her okay, keeping her the same as before, and she was spit out as if from a *Wizard of Oz* tornado. She felt like she and Cal were really alone.

She'd been wrong, of course: they'd met Sandy and Bo soon after. But maybe that was why Frida didn't like to think about that moment, because the Millers, who had seemed to be watching over them those first few months, weren't here anymore. Now she and Cal really were alone, and her old fears were too dangerous to revisit. Some feelings were hard to recover from.

She needed Cal. Her darling husband. She would call him in from his digging, tell him she was late, and he would remind her to breathe, and smile at her with his gentle, beautiful eyes.

She grabbed her hat and pushed open the door. Though it was overcast, there was still a glare, and she wished, yet again, for sunglasses. A breeze rustled the woods, and a far-off twig split from a branch.

Across the yard, Cal was pushing the shovel into the ground, his back to her. Behind him, the garden looked crowded and lush; the squash had come in, and once it was harvested they'd plant the lettuce and peas. The land had not given up on them, thank goodness. They had both been relieved when the rains came—and the house hadn't flooded. They had already lived through two winters here, and their third would be upon them soon. Frida would help Cal plant the garlic, if they could get it. If nature continued to cooperate, they would be okay.

Frida watched Cal push the shovel into the dirt and scoop it out. There were piles of dirt all around him, and the latest one was still small, the size of a science project volcano. Cal was muttering to himself, which meant he was worrying about something, unknotting some problem. She smiled and crouched behind the outdoor stove. She put her hands to her lips and whistled.

Cal lifted his head immediately. He looked past the crops to the line of trees there. Most were still green and lush, but some were starting to turn. Fall.

Frida whistled again, and Cal dropped the shovel. He was looking for a bird. She had fooled him. She saw him smile.

"Hello?" he called out.

Frida waited, her heart beating faster.

"Hello?" he said again.

Frida whistled back, *Hello, darling,* and this time Cal started. He slowly reached out his hand. Was it meant as an invitation? Did he think he was Saint Francis, that a bird would come to him?

She laughed and stood up.

"Fuck," he said when he saw her, and shook his head.

"I can't believe you fell for that." As she approached, she put her lips together and made the sound again.

"You got me. Good one."

She could tell she'd shaken him. "I'm sorry," she said. "Can I help?"

He shook his head. "No, but keep me company."

Frida nodded and sat down right on the dirt; it was cold, and she moved quickly to a kneeling position. She'd finally given in and worn one of Sandy's long dresses. It was made of denim and looked vaguely cultish, but it was comfortable and, with leggings beneath, warm.

She kept her eyes on the shovel.

"How deep do you need to go?"

He shrugged. "Deep enough."

She rolled her eyes. She hated when he offered vague, poetic answers to her questions.

"Sorry."

"I didn't get my period," she said. Why had she just blurted it out like that?

He looked at her carefully for a moment, as if willing himself to recognize her. "How late?"

"Too late. Thirteen days. You know I'm always on time."

On one wall of their home, Frida kept track of her cycle. She wrote with a chalky stone, sharpened to a point with the paring knife. She'd learned the system from Sandy Miller, who said she'd served as her own midwife for her two children. Frida liked the tallies and the circles, the order of it, how the body adhered to some invisible system. She sometimes called herself a hippie, told Cal she had an intimate relationship with the moon, but they both knew she took the record very seriously.

"Pregnant?" he said. He could barely get the word out.

"Maybe." She paused. "Or there's something wrong with me."

He nodded.

After they'd met the Millers, she and Cal had thought perhaps having children would be all right. Jane and Garrett breathed easily in this world and didn't want for anything, had no idea there was anything more to want. Maybe it was Frida and Cal's destiny to be parents. They even joked with Bo and Sandy about their families joining, as creepy as that sounded. Their children would mark the beginning of a new and better species, start the world over.

But Frida kept getting her period. And they made love all the time. Sometimes their lust was unquenchable, and sometimes they were just bored. Sex was the only fun, the only way to waste time. It replaced the Internet, reading, going out to dinner, shopping. The universe had righted itself, maybe. Still, no children. Now that the Millers were no longer around, Frida had begun to think it was for the best.

"So that's what's been bothering you," Cal said now.

She nodded. "Maybe it's just a nutrient I'm missing."

"Meat," he said, and nodded to his half-dug hole.

"I feel okay. I'm fine."

"You think August has a test?" he asked.

She laughed. "I doubt it. Eventually, I'll know one way or the other." She brought her hands to her stomach; it was still flat. "But maybe he knows a witch doctor. He could bring her over here."

This was a thing Frida liked to do: try and figure out where August traveled, and with whom else he traded. On his first visit he said he lived "around the way" and gave them a look that meant he didn't welcome personal questions. Short of following him, which they had promised never to do, there was no way of knowing where he went in the month they didn't see him. He refused to provide any clues.

He had once told Frida, "I'm warning you, don't be nosy. I don't serve the curious."

"When he comes," she said, "make yourself scarce. Go forage or something."

Cal thought about it for a moment. "He does like you better."

"And if I tell him our situation . . ."

"He'll at least give you a deal on the garlic."

They fell silent. Somewhere, far away, but not so far off, a bird began to call.

"What if you aren't sick?" he asked. "What if you're—"

"I can't even imagine it . . ."

She suddenly thought of her parents. Hilda and Dada, they called them. As if on cue, she thought of Micah, too. Dead five years.

"Hey," Cal said. "Don't go there."

She smiled, and he helped her to her feet. She took note of how careful he was being, how tightly he held her hand. Already, she realized, he thought of her differently.

Frida had imagined a child inside of her so many times, it was a wonder she had never given birth to one. She had felt her hips expanding, conjured morning sickness and swollen breasts, and sent love to an imagined fetus: fingerless and translucent, its heart glowing in its chest, tiny but there. Frida knew better and, in fact, often wished away the baby she had imagined. And maybe the wishing worked, because she never was actually pregnant.

Frida blamed Sandy for planting in her mind the notion that a family was a good idea. Not only that it could happen, but that it should. On one of their long foraging walks, Sandy had asked, "Who else will look after you in your old age?" as if it were assumed that Frida and Cal would live long enough to have that problem. Sandy believed strongly that the world wasn't going anywhere. The country was wrecked, yes, but something else, something better, something beautiful, was bound to replace it. Many times she had swept an arm through the air in front of her and said, "Look at what my children will inherit!" It wasn't hard to be seduced by Sandy Miller.

Frida had first met her down by the creek, which was only about a fifteen-minute hike from the shed. The walk was almost always pleasant; in the spring, Cal had pointed out the baby blue eyes and, in the summer, the clarkias. When she was alone, Frida would keep her eyes out for snakes, listen closely for other animals that might be hidden by the trees. That first summer, a porcupine had walked into her path, quills up, and Frida sucked in her breath and turned around, ran back to the shed crying like a kid. She imagined coming upon a bigger, more danger-ous animal, being eaten alive. Cal said she shouldn't worry, but he didn't call her crazy, either. They were in the wild, after all, and anything could happen—to think they could control their surroundings was foolish.

It had taken two weeks for Frida to find the bravery to venture into the forest by herself again. When she did, she was vigilant as a hunter and proud of herself for not turning back at every unknown rustle in the foliage. It only took two solo treks for her to get comfortable again. The green world filled her head and cleared it.

But on the day she met Sandy, their first summer out here was quickly turning into fall, and Frida's initial panic about their isolation had been replaced with a low hum of hopelessness. She barely saw the world around her. Not even five months into the afterlife, and she had turned to chores as a way to cope.

That day she was headed to the creek to wash some clothes. She pushed through the cattails to get to the edge of the water, the canvas bag of laundry bouncing over her shoulder. She should feel like a buffalo, she thought, heading to the water to drink. Instead, she was channeling a cartoon bandit or Santa Claus. As she stepped onto the muddy patch where she would do her soaking and wringing, something moved in the brush a few feet away. She froze. Between the grasses, a flash of corn silk. Barbie hair, Frida thought.

A very small boy popped up from the ground, and she let out a cry. She knew he was real because of the details she couldn't have made up: hair so blond it was almost white. The small scratches and bug bites on his arms and legs. Freckles all over his face, except his eyelids, which were as white as his hair. The man's T-shirt he wore like a dress, which read OFFICIAL PUSSY INSPECTOR. That made her laugh, and she glanced at his feet, which were bare, calloused into hooves.

"Don't be alarmed!" a female voice called out.

Frida looked up, and across the creek stood a woman, almost

as blond as the boy. She was tall and thin and wore overalls. It didn't look like she had on a shirt, but it couldn't have mattered—she looked as flat chested as a ballet dancer. A girl, older than the boy, with long brown hair, hid behind her mother's legs. She was wearing what looked like a burlap sack, sewn into a jumper.

"I come in peace," Frida yelled. What was this, some terrible alien flick? She started again. "Who are you?" All at once, she felt electric. They weren't alone!

Frida turned to the boy again and smiled. He widened his eyes, as if he, too, couldn't believe there was someone else to talk to. Later, Frida would learn that this was the way Garrett showed his pleasure.

Frida marveled at how quickly Sandy and Jane got across the creek. They knew which rocks would hold them and which ones were slippery with algae and should be avoided. Jane was barefoot like her brother, but Sandy wore hiking boots. Once across, she introduced herself and her two children. Jane was seven. Garrett was three. Up close, Sandy looked older: the face of a woman who worked outside without sunscreen.

"Bo and I," Sandy said, "we've been watching you two for some time. Making sure you were safe."

Frida nodded slowly. She and Cal hadn't counted on a family of spies. *Making sure you were safe.* Did that mean they were judging them, or protecting them?

"The birds!" Garrett cried. He was pointing at Frida now, as if he had just figured something out.

Frida raised an eyebrow.

"We took to calling you the birds," Sandy explained. "As in *lovebirds*. You two sure do like each other."

In a different context, Frida might have blushed. Instead she said, "It's cheaper than going to the movies." She was trying to keep her eyes off Sandy's chest. Her overalls had shifted in her commute across the water, and one breast, all nipple, peeped out from the bib, its tip long and knobby. It reminded Frida of a caterpillar.

"You're living in our shed," Sandy said. She didn't seem mad, and so instead of apologizing, Frida thanked her for building it. "I assume you don't want us to move out. It was empty when we arrived."

"Oh no, we love that you're there. It's where Bo and I first settled. We built that well you're using, you know. We like the little outdoor kitchen and fire pit you've added. I told Bo it was proof of your ingenuity."

"How long have you been here?" Frida asked.

"Forever," Sandy said.

That was the thing about the Millers: they never got specific. It was easy to deduce they'd arrived at least seven years before, since Jane was born on the land, but that was as much as Frida could figure out on her own. Sandy and Bo wouldn't say where they were from, either, though Los Angeles didn't seem to register much familiarity on their faces, nor did Cleveland, where Cal had been raised. It wasn't that their speech was accentless but that it shifted, from bland to twangy and back again in a single conversation. Once, Bo wore a Duke shirt, but Sandy said she'd gotten it from a friend, years and years ago. "Be protective of your past," she finally told Frida. "Our children don't need to know too much about ours."

On that first meeting, Sandy told her the names of the fish in the creek. "We don't know what that one's called," she said, pointing to a thin silvery one, "so we call it a princess." Frida wished she had a Device that worked, to take notes. She hadn't felt this happy in—maybe ever. Sandy's eyebrows were light as dandelion fuzz, and Frida loved the surprise of them. She hadn't realized how tired she'd gotten of Cal's face.

Sandy offered to help Frida with her laundry, and Frida accepted. Garrett ran up and down the creek, collecting rocks, and Jane stayed to help the women. Frida hadn't been taking much notice of her until Sandy said in a stern voice, "Hand that over." When Jane hesitated, Sandy snatched Cal's red bandanna out of her daughter's hand. She threw it to the ground as if it were on fire, her eyes squeezed shut.

"You okay?" Frida asked.

"She likes red," Sandy said. She affected a breezy laugh, but there was something shaky and nervous behind it. "We don't let her have too much of it."

"Sorry, Mama," Jane whispered.

The next time Garrett sped by, Frida tried to keep her voice casual. She didn't want to freak Sandy out again. "What's with his shirt?"

"He likes to help forage." Sandy raised an eyebrow, her eyes going twinkly for a moment. "Pussy is a kind of mushroom."

Frida laughed until she realized Jane was watching them.

"We find food, in the forest," the little girl said.

"Cool," Frida said. *Cool?* She supposed it didn't hurt, lying to the kids. It wasn't like Garrett would find out the truth. They could rename everything, if they wanted to.

The laundry was drying by the time Sandy led her kids away, back to their house. Frida practically ran to the shed. Sandy had invited them over for lunch the next day! The route to their house was easy, Sandy had said. With a stick she drew a rudimentary map in the mud. "And we've nailed hawk feathers into trees, to mark the trail. You haven't noticed them?" Frida shook her head.

It took some effort for Frida to convince Cal she wasn't playing a trick on him. And once he believed her, he was concerned. How did she know they weren't dangerous? Why had they brought children into this world? "That's troubling to me," he said, but Frida wasn't eager to follow this line of argument. He sounded like her brother when he talked that way—all doom.

"I'm going whether you come or not," Frida had said, and that settled it.

The Millers' house would have been impossible to find, were it not for those feathers, and those key phrases chiseled into Frida's brain: "Turn left at the boulder, walk until you reach two fallen trees, one atop the other, forming a cross. Turn right." A few times, Frida felt a flash of nervousness that they were lost, but an hour later they pulled back a large branch, attached to which was another feather, tied with turquoise-colored leather, and entered a clearing. Across the field, a house materialized. Frida felt victorious.

Compared with the shed, the Millers' home was enormous, and durable, its exterior built of stone and wood. The family must have heard them approaching because all four of them were waiting outside the front door.

"Are they getting their portrait done?" Cal whispered. Frida barely registered the comment, so transfixed was she by Bo's naked face, no beard to obscure it. Cal himself had a thick beard going, the same look Micah had sported when he left for college, as if he hadn't been raised in a city, as if he'd ever gone camping. She kind of liked Cal's copper-colored beard, but maybe this Bo could teach her husband how to shave with a knife. What she missed was having the option.

"Welcome." Bo stepped forward and shook hands with both of them. He did not smile. He was shorter than Sandy but sturdy with muscles, barbed with them. His seriousness took something away from him, Frida thought, his high cheekbones and heavy black eyebrows menacing rather than dignified. And he squinted, as if he'd lost his glasses. Perhaps this was a man who had been broken down by blurriness.

"We're so happy you made it!" Sandy said. She wore the same overalls but, thankfully, had added a blue T-shirt to the ensemble. She held Jane's hand, and Garrett was slung on her hip. At Frida's greeting, the boy rubbed his left eye with a fist and shook his head. "He just woke up from a nap," Sandy said. The little girl nodded, as if confirming her mother's story.

Bo invited them inside, and Jane skipped forward to lead the parade. The house was one large, low-ceilinged room, with two cubbylike spaces for bedrooms. They slept in real beds; Sandy and Bo's had a wooden headboard, and the children slept on what looked like sturdy cots. Frida saw Cal take in these comforts. In the shed, she and Cal had four sleeping bags, which they rotated or layered. No pillows.

"What a place," Cal said. Later, when he was trying to make Frida laugh, he would refer to it as the Miller Estate.

There were no windows, so the house was dark, but it was surprisingly cool, like a basement. They could open the door for light and air, Sandy explained.

In the middle of the room two mismatched couches faced each other. The setup reminded Frida of a rundown rec center or a home for the elderly gone sadder than expected. Someone had built smaller chairs for the kids, but they looked about as comfortable as birds' nests: twiggy and sharp. On the rudimentary wooden table nearby, Frida counted two oil lanterns and half a dozen candles.

With Garrett still on her hip, Sandy moved toward the kitchen area. It was just a stone fire pit, and a trashed card table. No chairs. Bo had built shelving into the walls, and on these the family's dishes and tools were crowded. Frida took note of

23

the plastic tarps, folded on the bottom shelf. She wondered if the house leaked.

"We do most of our cooking outside, or we eat raw," Bo said. "Smoke from the fire pit in here won't kill us—there's a chimney of sorts—but we could've designed it better."

Sandy smiled. "I hope you're hungry. We've got rabbit, just need to put it on the fire."

Frida squeezed Cal's hand; she couldn't remember the last time they'd eaten meat.

"We use snares," Bo said, and Cal said he'd love to learn more.

Bo offered to show them the root cellar next and the outhouse and their new underground shed, where they were doing their curing.

Despite his initial austerity, Bo treated Frida and Cal with a tenderness that seemed Southern. He often used their names when speaking to them, as if his conversation were a gift. "You see, Calvin," he would say, "snares can be difficult to build, but they're quite efficient." Like his wife, Bo wore a gold band on his left ring finger. So they'd been out here awhile, Frida thought, long before the world really went to shit. Hilda and Dada had given Frida their rings as a wedding present, but she and Cal had sold them not long after.

"You two married?" Sandy had asked her at the creek. No wonder.

With the Millers, Frida felt like she'd fallen asleep and awoken in a bygone era. They could have been pioneers, hitching their covered wagons, staking claim on a new frontier. Manifest destiny bullshit. Or the opposite: with Bo and Sandy, the land outside wasn't wild and uncharted, something to fear until conquered. No, the earth was to be respected. Only then would it collaborate with you, tell you what it needed and what it was willing to give. And it was willing to give you a lot, if you knew how to ask. It was a lesson in coaxing.

After they'd eaten a meal so succulent and satisfying Frida could have moaned with pleasure, Sandy asked her to follow her back into the house. The men had begun to discuss how to handle larger predators and keep the deer away from food storage and scare off the rare bear that skulked the grounds. Bo had once seen a mama bear and her cubs at the edge of the land; "Imagine if I'd been near them," he was telling Cal. "They're just animals and I've got a gun, but still, I'm not stupid. They scare me." It was a conversation Frida thought she should be involved in, but what the hell, she could get a distilled version from Cal on the walk back to the shed. She wished Sandy and Bo would invite them to stay over, but she knew they wouldn't. Already, Bo had made it clear that they would not be seeing one another all the time. "There's always work to be done," he'd said during lunch.

Sandy had grabbed Frida's hand as they walked into the house. It was as dry as Frida's own, her knuckles white and flaky. "I guess you won't be lending me any lotion," she said, nodding at their intertwined hands.

"I wish. I'm dry as an old lake bed. But I did want to show you this."

They were like two little girls on a playdate, like Sandy was about to reveal her secret doll collection, her stickers, or her mother's lacy lingerie. Jane tried to follow them inside, but once they were a few feet into the house, Sandy had turned around and said, "Go to Papa."

Once Jane was gone, Sandy pointed to the far wall, just to the left of the bed she shared with Bo. Frida had seen the grayish marks earlier, but had taken them to be Jane's scribbles: the cave paintings of a seven-year-old.

"Go ahead," Sandy said, and Frida let go of her hand to walk closer.

Of course the drawings couldn't be Jane's, they were too far up the wall. At the top, a line of carefully drawn circles, some of them shaded in, others only partially.

"The phases of the moon," Sandy said behind her, and Frida raised an eyebrow. She hoped Sandy wasn't inviting her into a coven.

"You can't just run to the store for tampons," Sandy said, and Frida understood what this calendar kept track of.

"I figured that out pretty quickly," Frida said. She didn't bother to tell Sandy that most stores in L.A. had found the needs of women harder and harder to meet.

"You can't be teenagers forever," Sandy said. "Cal should give you a child."

"Excuse me?" Frida said. No wonder Sandy had made Jane stay outside. "I don't think I understand what you mean."

"Yes, you do," Sandy said. "Lovebirds. Eventually there's a cloacal kiss."

How close had the Millers been watching them? Close enough. They had seen Cal move off of her, just before he came. She and Cal liked to do it outside, if the weather was nice. Frida wanted to sew this strange woman's mouth shut— or, better, her eyes.

"I don't think that's any of your business," Frida said.

Their birth control of choice was common back home. She didn't know anyone who did it otherwise; it wasn't foolproof, but no one she'd known had ever had an accident. And, thank God: Who wanted to bring children into this world? Who could find a doctor, who could afford condoms, let alone the Pill?

When Frida was in high school, she'd taken it to help ease

her cramps. She'd loved the little pink clamshell they came in and the way the tiny tablets popped out of their plastic sheaths. But before her senior year began, Dada started having trouble finding work, and gas prices were rising every week, and the family began its Great Austerity Measures, as Hilda put it. Goodbye clamshell and a menstrual cycle Frida actually kept track of. Goodbye almost everything frivolous and easy.

By the time she and Cal had agreed to leave L.A., it seemed like no one had access to meds; only the deranged would buy a handful of drugs from a guy on the street corner. Was that really Xanax wrapped in tinfoil? Prescriptions, like doctors, were for the rich. The lucky ones, the people with money, had long fled L.A.

"I apologize if I'm embarrassing you," Sandy said then. "I didn't mean to see."

"Don't you believe in privacy?"

"Not really, I guess."

Frida didn't know what to do with Sandy's candor. She finally asked: "Why are you showing me this?"

"Because it's your responsibility. It's everything," Sandy said.

In the doorway, the sun caught the lightness of her hair, and it seemed for a moment as if she wore a halo.

"Don't tell me you came out here to die."

Frida was about to ask Sandy if she was nuts. She wanted to say it was too risky to have a kid, that it was selfish. What if they got sick? What if there wasn't enough food? What if, what if. But Sandy was already turning around. She left Frida alone in the dark house.

Cal admitted he'd been wrong, that—after spending the afternoon at the Millers' place—he trusted them. "They have small children," he said that night, once they'd finally reached the shed, just before sunset, thank goodness. As if he hadn't known about Jane and Garrett before he'd met them. As if people with small children couldn't cause harm. Frida decided not to tell him what Sandy had said. They would be seeing the family fairly regularly, and as weird as they were, Frida was relieved they existed.

"But I do wonder where they get the salt to cure their meat," Cal had added. Frida didn't have an answer, and, anyway, it was the farthest thing from her mind, and she didn't press Cal to go on. She couldn't stop thinking about what Sandy had told her in the house. It changed things. Frida felt her perspective shifting, tilting the world, blurring the colors, brightening them.

The next time they had sex, when Cal said, "I'm close," Frida held him to her, wouldn't let him go. "Good," she'd whispered into his ear.

They didn't talk about what had happened, not at first. When they did, they both admitted it felt right. Having the Millers nearby, just the very idea of them, gave them both solace. The hopelessness lifted right off of Frida.

Three weeks later, the Millers arrived at the shed. Already Garrett looked older, taller, and someone had given Jane a bob.

"You look like a flapper," Frida had told her that day. The girl frowned. Of course she had no idea what that was.

"It's a kind of lady, from a long time ago." Jane waited, as if expecting more, and Frida kept talking. "From like a hundred years ago ... actually longer, maybe close to a hundred and thirty. A long time." Frida paused. "She liked to dance."

At this Jane beamed, but a moment later, as if startled by her own joy, she turned away from Frida, hiding her face in her mother's thighs. Sandy said, "Sometimes Garrett bangs on the drum we have, and Janey dances."

Frida laughed, and so did Sandy.

"Do you mind showing me the shed?" Sandy asked. "I'm curious to see what you've done with the place." Frida agreed, and Sandy grabbed Jane's hand. The three headed to the shed.

When they reached the open doorway, Sandy looked up, her eyes on the dark interior. Suddenly, she stepped back into the sunlight and pulled Jane's hand so roughly her daughter crashed against her thigh. What was it? Cal's bandanna wasn't in sight, but then Frida saw it: her sleeping bag was a bright red.

"You okay?" Frida asked.

Sandy said nothing, only stepped farther away from the shed, dragging Jane with her.

"Sandy," Frida called out, but Sandy was already halfway to the garden, where Bo, Garrett, and Cal were bending over something in the dirt.

Frida followed them. When Sandy saw Frida behind her, she forced a smile and said, "Oh! I almost forgot. We brought you some stuff."

The Millers had come bearing gifts. A rabbit, already skinned and ready to roast. Also some chanterelles. "Sandy will show you how to find those," Bo said to Frida. The subtext being: *I hunt. You, Woman, shall gather.*

The third gift was the most surprising. Sandy smiled at her, as if to say, *Let's forget about what happened in the shed,* and pulled from their bag a box of Band-Aids. Frida yanked it out of her hand.

"Where the fuck did you get this?" she asked.

"Frida, calm down," Cal said, but Bo was laughing. In another minute, Sandy was, too; she seemed totally relaxed now. Frida felt relieved.

"It's okay," Sandy said. "They are exotic, aren't they."

Frida flipped open the tin's lid. Inside, the Band-Aids behaved so well, lined up like schoolchildren. Already she was imagining plucking one out. Its white wrapper thin as rice paper, and those tiny blue arrows at the top. Open here. How it would peel back so easily to reveal the Band-Aid itself, nestled flat inside. Frida's stomach fluttered. She could have sucked on it. The salty, pretzel taste of wounds.

"Thank you," Frida said finally. "How long have you had these?"

"A few weeks," Bo said. "We traded for them."

That's how they learned about August.

"He travels widely," Bo said. "He won't tell us how many others are out there, but there are a few."

"Is that so?" Cal said. "I guess this is the place to be. Who knew that—"

"Don't," Bo said, holding up his hand.

"Don't what?" Frida asked.

Sandy smiled weakly. "Never say where we are."

"It's something we decided on," Bo said. "The state. Place-names. Keep all that out."

Sandy added, "It feels more private this way."

"I thought you didn't believe in privacy," Frida said.

"You got me there."

The men didn't catch their little joke. They were clueless. Some things didn't change.

"Think of it as a place of mystery," Bo said.

Later, Cal said the Millers were a little nuts. But he liked the rule. "This place of mystery," he said. "It's got a ring to it, don't you think?"

Yes, Frida had to admit. It did.

2

Cal had to hold himself back from touching Frida's hair as she slept. It lay dark and wavy across the pillow, shinier than the creek at midday. She sighed and turned over, pushing her heel into his calf so hard it almost hurt. Asleep, her mouth a flat thin line, she looked plainer than she was; without her big soulful eyes—how they lit up when she smiled or pried into him when she was serious or upset—her beauty was evident but unremarkable. She looked older than twenty-nine.

The sun hadn't yet risen, and here he was, staring at his wife, wide awake. He'd give his left nut for a new book or the chatter of a sports-radio jackass. Anything to get through this almost-dawn. At least it wasn't that cold; someday they might be trapped in here because of a freak snowstorm. And then what?

There was bound to be some bad weather down the line: a drought or relentless, heavy rains. They'd wanted to avoid the kind of storms that had battered the Northwest the year before they left L.A.; those states had barely recovered, and he and Frida had settled here because the climate was milder. So far, they'd been spared really bad weather. But for how long? He turned the question over and over in his mind.

This was one of things he loved about life out here. The space to consider questions. Even if he sometimes longed for mindless diversions, mostly he was grateful for the silence, the time. It reminded him of college, where thinking itself was considered noble, and where there had been nothing to distract from that endeavor. For most of his fellow students, it had been the first break from Devices, but Cal had never owned such things. There were too many links to cancer, his mother said, and she wanted him to feel lost once in a while. Everyone else was dependent on instant answers, on satellites, and this was turning them stupid. He'd written papers about how painful the shift to digital had been, about people's addiction to the Internet, about how the batteries in dead laptops were leaking into the earth. That last one had been for the Politics of Geography course his mom had designed. Cal had been homeschooled. His mother had taught him everything he knew, until he took the train to a tiny town in California and started college at Plank.

There, they'd stay up all night, considering the nature of being, of meaning. In retrospect, they seemed like such stoners, holding up a flashlight and questioning their perception of it. If they weren't Americans, they asked, how would they see it? If they weren't men, if they weren't privileged . . . the questions were endless. Most of the time they weren't high, just serious. Too serious, probably. But they had joked around a lot, too. Especially Micah.

The first thing Cal remembered about Micah was how he sat on the edge of the bare mattress in their dorm room, slumped over like an old man sleeping on the bus. But he wasn't sleeping, he was reading a small worn paperback. A few years prior, the student body had voted against allowing e-readers on campus—previously the only electronics allowed—which made it almost impossible to access contemporary books. Not that it mattered. On their first date, Cal had told Frida that Plank loved D.W.M.: *dead white men.*

Cal never caught the name of the book Micah was reading that first day, because as soon as Cal said hello, tentatively, lugging his suitcase into the high-ceilinged room, Micah had jumped from his pose and tossed the book aside. It fell between the bed and the wall.

"You're here!" Micah said. He rushed forward to help Cal with his bags. He was taller than Cal and almost burly. He had a beard, which was rust colored in places, though his hair was dark. He looked older than eighteen. Cal figured he must be from a place like Montana or Maine. Definitely not a city.

"Micah Ellis," he said, offering his hand.

"Cal. Cal Friedman."

"You're Jewish?"

"My mother is. Was. She's an atheist."

"You took her last name?"

Cal nodded. "But I see my father, all the time."

"No judgment." Micah held up his hands. "My parents are married, and we don't practice any religion. Where you from? Did you say? Are you eighteen?" He smiled, almost sheepishly. "I apologize for the questions, I'm somewhat of a taxonomist."

At first, Cal thought he'd said "taxidermist." He pictured this bearded kid in a basement in Maine, stuffing bobcats and bears and, then, Cal himself.

"I like to classify things," Micah was saying now, and Cal understood that he had misheard his new roommate. "Where you from?"

"Cleveland."

Micah grinned. "I could have guessed from your accent. Flat, nasal."

Cal knew he should be offended, but he wasn't. "And you?"

"L.A."

"Really? That's hard to believe."

33

"Why? 'Cause my tits are real?"

They both laughed.

Micah had arrived a couple of hours before. The room was large, and at that hour sunlight pooled through the circular window between their dressers. Even now, Cal remembered how golden the light had been at Plank. In the morning, the sun spread across the floor and his desk—he kept his neat, almost bare, whereas Micah's was always covered with books and pens and dirty dishes. In Cleveland, Cal and his mother used blackout curtains, but at Plank the windows were naked, and he often woke at dawn even if he didn't have to get up then to work the school's small farm. A previous generation of Plankers had probably voted against drapes, in the same way they had rejected the Internet and the coed question, the gingham curtains burned in a bonfire one crisp winter night, the boys howling.

"I like the room," Cal said, after they'd agreed who would sleep where. "I feel like I'm in a time machine." What he meant to say was: Plank felt lost in the past. Not stuck, but suspended there, in its beauty and slowness.

"We're encased in amber," Micah had said, and smiled.

Plank's student body was made up of thirty male students. All of them lived in a converted farmhouse, though two second-years got to board in the house's former kitchen, coveted for its wood-burning stove. Cal was the only first-year from the Midwest, and one of the few kids who hadn't gone to a prep school. Micah hadn't either, but he'd attended an intense public high school in L.A. where you had to test highly gifted to get in. It closed from lack of funding a year after he graduated. Micah couldn't milk a cow, as Cal could, but he had already read Plato and Derrida. "The jugness of the jug" was how he explained Heidegger to Cal, as if that explained anything at all.

It made him laugh now, thinking about the way Plankers used to talk. They'd farm in the mornings, bring the goats out to pasture, and then, with dirt under their fingernails and smelling of animal shit, they'd head into seminar to toss big words back and forth at one another.

Out here, in the middle of nowhere, the real middle of nowhere, those big ideas offered him solace but not much else. Cal glanced once more at Frida, whose eyes were shut tight against the world, and he wondered what she might say about all the books he and her brother had once devoured. *Like any of it could rescue you,* she might say.

There were only two years at Plank. If you were admitted, it was free, but there was no real degree at the end. Most of the boys transferred to one of the Ivy Leagues, went traveling, or fell off the map. Cal's dad had been the one to show him the application. He ran his small organic farm holistically, which meant the cows were moved often, so as not to wreck the land, and the chickens followed, pecking at the manure, and the vegetables were grown without pesticides. Cal's dad had always urged his son to learn his trade. "You have skills that this school will nurture." But did his dad know that beyond working the school's alfalfa farm, milking its cows, and learning to slaughter the occasional goat, Plank's students trafficked in the abstract?

Many Plankers wanted to fight injustice and poverty throughout the world, though certainly not with religion; it seemed like everyone was an atheist or headed there. They'd use their brilliance and tenacity, not God, to make a difference. His first week, Cal heard another student talking about his plans after graduation; the guy had a whole business plan already written, but he wanted to get everyone's opinions on environmental tariffs and microloans. Cal had never met anyone like that before, a person so open about his ambitions, but at Plank it was common for someone to announce his lofty ideals over a meal or in class or in the lounge at 3:00 a.m. Plankers would change the world. A lot of other colleges had closed in recent years, but Plank was cheap to run, and its endowment was solid because its alumni believed their small but mighty network of graduates would solve the crises that blighted the present. He hadn't realized it when he was accepted, but after he'd arrived, it was clear: Plank expected something of him. He was not to take his education in vain.

Cal remembered how, on that first night, the older students cooked and served dinner to the new ones. The second-years had baked bread and cooked a spicy vegetable soup. They wore aprons and belched with abandon. On the walls of the large dining room hung old farming equipment: a hoe, a rusted pickax. Otherwise, the large room was bare and dingy. Already, the older students had made it known that nobody much cared for how things looked. For instance, one of the windows had been shattered in a pickup football game, and instead of replacing the glass, as the president wanted, the boys had voted to duct-tape the damage.

All of it had been a lesson, Cal thought now. He'd taken none of it for granted. His time at Plank had prepared him for the devastation in

L.A. and their life out here. He knew not only how to skin an animal and how to irrigate a field but also how to forgive a room for its ugliness. Frida was sensitive to space; she said the Millers' place—their place now—was so utilitarian it was like living in a police station. He hadn't even noticed how ugly the shelves were until she'd pointed them out. Frida probably thought this blindness a flaw, but he considered it a skill.

Like right now, Cal thought. Frida found the house depressing and dank, and she hated their practical yet dumpy couches.

But what he saw, what he felt, was different. Here he was, lying next to Frida, whose body was warm and solid next to his own, and they were okay, they had shelter, they had each other. The other particulars, the lack of windows, the slanting shelves, didn't matter.

Frida pushed once more into his calf. She muttered something—it almost sounded like another language, harsh, with a lot of consonants—and then she fell back into a silent sleep. She looked calm and comfortable in this bed, in this life.

Cal remembered that first dinner at Plank, how the other boys used words he'd never heard of: *enframed, signifier, telos,* and *phronesis.* What he'd learned about the world so far was baby food compared with what these guys knew, and that night he took to nodding at the things his classmates said while inside his brain, a tumbleweed skipped. At orientation before the meal, the second-years had taught them to show agreement by raising their fists and knocking on an invisible door. Cal couldn't see himself ever doing this, not without laughing, at least, but at dinner, there was Micah with his fist up, knocking. As if he'd always known the gesture.

Micah and another first-year were discussing a German film they both liked. It was about terrorists from the 1970s. Micah's great-grandmother remembered the nightly news reporting their violence when she was young, although she couldn't keep track of what they'd opposed. "She couldn't even remember what war everyone had protested back then," Micah said. "I told her, 'Vietnam, Grammie,' and her eyes lit up, like she'd won a prize. No joke." Then he mentioned some artist's rendering of the German terrorists. "The portraits hit me in my core with a hot poker," he said.

"How so?" Cal had asked. He supposed he wanted to try to participate in the conversation, in this weird little world; it had to happen eventually, or he'd go nuts. But afterward he wished he had left Micah be, that he hadn't said anything. Or he wished he didn't remember the moment now, almost a decade later.

The sun was about to rise, Cal could feel it. Maybe Frida felt it, too; maybe in her sleep, the day was calling her forth. But Micah's answer was still there, in Cal's head.

"'How so'?" Micah echoed. "The pictures are fucking brilliant. Those terrorists are rendered mysterious and grim. That dark gray blurriness . . . they're painted from actual photographs, you know that, right? Part of me always wishes they weren't blurry, but that's what makes them magnetic. And even as I'm magnetized, I feel a dispassion. Sure, that's a dead body, you might say, but unless you know the history, the context, does it even matter?"

Cal was too embarrassed to admit that he'd never heard of the artist or seen the paintings, so he just nodded and waited for someone else to say something.

In that space, Micah spoke again. This time, his voice was gentle, softer, almost like he was waking a sleeping child. "Violence is beautiful, in a way."

He smiled a big lusty grin, and Cal felt like he'd been socked in the stomach.

"What is it, California?" Micah said. "Don't you agree with me?"

Cal had been so stunned by Micah's answer he barely registered the nickname.

"That's your full name?" a kid across the table said. "California?"

"Awesome!" someone else said.

"Isn't it though?" Micah said. "His mother's a hippie."

"Hippies don't exist anymore," another Planker said.

38

"Apparently they do," Micah said.

"No," Cal heard himself say. "It's Calvin. My name is Calvin."

At this, Micah groaned. "God, Cal, we almost had them! Couldn't you go along with it?" Then he roared with laughter, and someone else said, "Fuck off, man, fuck off," laughing, too, knocking his fist. Cal didn't say a thing. Back home, he'd never been the talkative type, but he wasn't shy. At Plank, he was quickly earning a reputation for being reticent and thoughtful. What a fraud.

After that, people sometimes called him California. It was a female name, it seemed, for the guys said it to him sweetly, like they were talking to a beautiful woman in a dark jukebox bar. Someone needed to be the girl, if only for a moment.

Micah's little trick with his name, Cal thought, was proof that he was a liar.

A little light began to seep under the front door. In no time the whole thing would be framed with sunlight. Bo hadn't built the best door, and in the winter they had to hang a thick rug over it to keep out the cold and, at the height of the summer, netting to keep out flies. These few weeks were the only time the door worked just as it was, letting in a breeze and those rays of sunlight.

Frida rolled back into him and grumbled more dream-babble into his neck. She slept so deeply, she was probably in L.A. now, ordering a latte. That dream often teased her, and she awoke upset. At least she wasn't having the nightmares anymore, the ones about Micah. "Him leaving" was how she put it.

At his funeral, someone should have included Micah's prankster nature in a eulogy, but no one did, no surprise. Not that his lying was a bad thing. Micah never carried a lie for long—what he enjoyed most was revealing his trickery.

Why was Cal thinking about Micah? About Plank? Time moved forward, but the mind was restless and stubborn, and it skipped to wherever it pleased, often to the past: backward, always backward. He wished he had an empty journal to scribble in. If he did, he'd get all this down. But he needed the pages he had left for practical purposes.

He looked at Frida once more. Her face was calm and blank, her mouth open now, the same expression she used to make when putting on mascara. Lines had begun to form around her eyes, and he was happy their only mirror was the rearview, taken from their car at the last moment. If they had a better one, she'd certainly complain about the smallest wrinkle, hold her face back with her hands, as his mother had done. "My face-lift," she'd say.

If Frida really was pregnant, what would they do? He imagined cutting the umbilical cord with his Swiss army knife. He knew Frida would want to go find others; they weren't the Millers, she'd say, and that was true. Families couldn't exist in a vacuum, or not theirs, at least.

Or could they?

He felt her hand on his thigh.

"Hey you," she whispered. For a second Cal wondered if she'd been faking sleep all this time.

"I couldn't sleep," he said.

"Poor thing."

"Any lattes last night?" he asked.

She shook her head. "Ugh. Don't dangle that in front of me. Not this morning."

Frida yawned and sat up. She was wearing the oversize T-shirt she preferred to sleep in until it got too cold. The shirt had once been white and was delicate as tissue. A small hole had opened at her right shoulder blade, as if her bone had been sharp enough to rip through the fabric. Cal poked his finger through and touched her skin. Cal figured the shirt had once belonged to her father, but he had never asked. To bring it up now would only rattle her. She'd been wearing it since they'd starting sleeping together, and he loved how her nipples showed through the front.

Frida yawned once more and climbed onto his stomach, so that she was straddling him.

"You look good," Cal said.

"Seriously?"

"Seriously."

He reached for her breasts, cupped one in each palm. She smiled.

"Good morning to you," she said. She nodded to his crotch. "And to *you*."

He laughed, running his hands to her waist. "You know what I like best about this place?"

She frowned. "What?"

"No one can hear me fucking your brains out."

Frida blushed. He wanted her so badly. He loved that he could say this to her, that she wanted him to say it, and that nothing had ever felt so natural.

She had her hand on him now. She leaned forward, and he could smell her musky breath. "Sorry," she said. "Morning breath?"

He grabbed her jaw and kissed her. "Yum," he said, and pulled for her T-shirt.

Afterward, once they were lying side by side, Cal got to thinking again of that first night at Plank, about Micah. The beauty of violence, all that nonsense, Micah's grave voice, as if he were imparting something vital to his new roommate.

Stupid Cal, he thought. Get that out of your head.

"Helloooo?" Frida was saying. With two fingers she flicked his dick lightly.

"Sorry," Cal said.

"What's wrong?"

"It's nothing . . . I just can't get a painting out of my head."

He couldn't tell her much more. After he'd graduated and was living in L.A., he'd tried searching for the pictures online. There had been one artist, near the end of the last century, who drew from photographs. Cal couldn't remember his name; he didn't want to remember it. The portraits were gray and deliberately blurry, just as Micah had described them. They were haunting, but they weren't magnetic or beautiful. Not to him.

"A painting?" Frida said.

"Yeah," he said. "More than one. They're of "—he didn't want to say the word, but he had to; there wasn't a way around it—"they're of terrorists. From Germany. In the 1970s." He paused. "We were into them at Plank," he said.

He thought his voice sounded innocent, like he was just talking about some random artist, but Frida sat up immediately. She was pulling on her T-shirt.

"Why are you thinking about *that?*" she said.

So she knew he was talking about Micah. Micah must have talked about those paintings with her, too.

They tried not to bring up her brother. They'd agreed to never tell anyone out here about him, and even between them, both his death and his life were difficult subjects, so thorny they could cut themselves on his name.

If Cal told her Micah had liked the paintings back when they were at Plank, she'd freak out. In the world according to Frida, her brother had been a precocious boy and then a brilliant man, faultless until the Group got ahold of him. According to her,

42

Micah never would have said such a thing before he became involved with them.

"It's nothing," Cal said now. He tried to pull Frida back, but she was already slipping out of his grasp. She crawled over him to get out of the bed. He couldn't help but look at her nipples. There they were.

"Babe," he said, sitting up. "Why are you so mad?"

"Please, don't," she said. She put on her pants.

"I'm going to the well," she said. She was already pulling on a sweater, her boots.

"Frida—"

But she was opening the door, the morning light spilling into the house, falling across all its dusty surfaces, its sad furniture. It was ugly; Frida had been right.

Cal called her name, but only after the door had slammed behind her.

3

Frida was chopping beets for dinner when she heard the crack of twigs and the rustle of trees and, after that, an animal's hooves against the hard dirt of the path to their house. She could just barely make out August's whistling, and then he stopped to mutter something. It sounded soothing, and she knew he was talking to the animal, Sue, probably congratulating her on another safe arrival.

Frida put down her knife and turned to Cal. They hadn't spoken much beyond the necessary all day. She wasn't sure why she was mad, and at whom, even. Her husband had brought up something stupid, something from a long time ago. It shouldn't matter now, but it did.

They didn't talk about Micah because, when they did, Cal got pissed and Frida got sad, and everything miserable about the world wedged itself between them. Frida knew what her brother had done; she had accepted it; she wasn't in denial. But sometimes, the Micah that Cal remembered had not a lick of goodness in him. That dumb college kid who had once been into Gerhard Richter and other pretentious shit? That was her *little brother*. Whatever else he was, well, it didn't erase that fact.

Cal probably wouldn't bring up her family now, not for weeks. He'd be afraid that if he did, she'd fall into a grief spiral. As if never mentioning her brother or her parents made her longing for them disappear. As if he could will all that pain away.

Cal was sprawled across one of the couches, an arm slung over his eyes. He was thinking. This was an actual activity they did now—just lay down and let their minds wander. Sometimes they gave each other a term to meditate on: *Magic Marker, air conditioner, strawberry.* It was more entertaining than Frida would have ever imagined it to be. Sandy Miller had told her about it. She and Bo used to do it, before they had kids. "Jane and Garrett keep us busy," she'd said. That made sense now: parenting as a way to kill time.

"He's here," Frida said, and Cal moved his arm off his face, sat up. "Get your foraging gear and meet him on your way out."

"You can hear him? Your ears are as good as a dog's." He grinned, his version of an olive branch.

"Hurry," she said.

Frida dried her hands and placed the beets in one of the metal bowls. With a little dried mint, they would be all right for dinner.

She heard August greet Cal. After a few moments, Cal yelled her name, and she headed outside.

The two men stood at the edge of the clearing, Cal with the foraging bag over his shoulder, the gardening gloves and paring knife in his hands. August had jumped off his buggy and was running a brush across the mare, who snorted at his touch.

The first time Frida had seen him approach the shed, sitting high on his carriage like someone out of Victorian England, she had felt oddly homesick. The carriage, choked with discarded furniture, car parts, crates of produce, and even a dollhouse, reminded her of those rundown trucks in L.A., filled with junk. There was always a hand-painted phone number on the side, to call if you needed something picked up and discarded. When she was younger, it had been a job for illegal immigrants, but over the years, more businesses like it began popping up, with all kinds of drivers. Near the end, they'd begun to disappear; you had to have a safe place to store your truck and its discards, or else all of it would be looted, and almost no one had that.

When she told August about these trucks, he had shrugged. "I've been out here a long time," he said. But what was a long time? She'd wondered if he'd struck out for the wilderness before the earthquakes. At the time, Frida had been seventeen, Micah fifteen, and

L.A. never recovered from the destruction. Nor had San Francisco, six months later. In the year following, the film industry—the kind that paid Dada, at least—left L.A. altogether, and the rich fled to the new Communities popping up everywhere. Hilda took to crying a lot and saying, "What now? What now?"

If August hadn't seen the reports of wildfires in Colorado and Utah or, later, those snowstorms across the Midwest and the East Coast or the rainstorms north of here, he would have no idea how battered the world was. And besides, would they have bothered a man who only whispered his secrets to a mule?

August was wearing what he always wore: a gray sweatshirt and sweatpants, the pants pushed to his calves like britches; white tube socks; and the black lace-up boots of a soldier. His head was covered, as always, with a black beanie, and his wraparound sunglasses shielded his eyes. He never took them off. Frida hated how she saw herself in their reflection, which kept her from looking him in the eyes. His intention, she presumed.

He nodded at Frida but returned his gaze to Cal, who had begun walking backward.

"I gotta run while the sun's still up," Cal was saying. "She'll take care of everything."

"I'd expect that," August said.

Frida was close enough now that she could greet him properly. Always a handshake.

"Nice to see you," he said. "You look well."

"Thank you," she said. "I feel great."

Cal looked away; he had the worst poker face.

Once Cal was gone, August invited Frida to come around to the back of the carriage. "I have a few new things," he said, and climbed up. Frida remained where she was. No one was allowed up on the carriage except August.

"Do you have any garlic?" she asked. "Cal wants to plant some. For flavoring, obviously. But also to ward off colds." This might be a perfect segue, she thought. Something about how she'd need to stay healthy, that the stakes were higher now that she might be pregnant.

"Let me check," August said, and rummaged through his belongings, which, today, included an old bicycle, missing its seat, and a pile of tarps, one of them already shredded to confetti. A moment later, August was grinning. "I've got Vicodin."

Had she heard him right? She'd never been much of a pill popper—as a teenager she'd preferred weed above all else—but she imagined the Vicodin sliding down her throat, on its way to making her feel good. A buzz: that's what she wanted.

"Did I hear you right?"

"I knew that'd get your attention. Always took you for a party girl." August pulled something from a mesh bag and stuffed it into his pocket. He climbed out of the carriage. "I've got a couple of the big boys. Seven hundred fifty milligrams."

Frida nodded. If she was pregnant—what would happen? "I thought the Communities had killed the drug trade." She remembered reading about it back in L.A.; the Communities were so safe and clean, even smoking a cigarette could get you exiled. That, and not paying your membership fees. "But I guess they've got to have a black market."

August just raised an eyebrow; he never wanted to talk about the world beyond.

"I guess Vicodin was always legal with a prescription," Frida said, keeping her eyes on him. "And those Community bastards still have access to everything that makes you feel better. Have a cold, call the doctor, et cetera, et cetera. Right?"

August was silent.

"What are you asking for it?" Frida asked finally.

"I knew it," he said. "You love pills."

"I was always more slacker stoner than glamorous party girl. A pothead through and through." She shrugged. "But I could use a little fun."

"But you don't have any pain," he asked, "do you?"

"Define pain," she said, and laughed. But he remained serious, and she shook her head. "I told you, it sounds like fun."

He said the pills would cost her. After a couple of offers, he finally accepted a bra, barely worn when they had moved here and almost forgotten. Frida knew Cal would never notice its absence.

"I'll throw in the garlic for free," August said, reaching into the carriage to stuff the bra into a duffel bag. "The bulbs I've got are a little shriveled, don't know if they'll take anyway."

"A steal!" Frida cried. She wouldn't have to tell Cal a thing.

From his pocket, August pulled an amber-colored plastic vial, the prescription information torn off. "Put out your hand," he said. He shook out two white pills onto her palm.

"Two seems a bit much," she said.

He handed her a canteen. "It's water," he said, and Frida threw one of the pills back before she could change her mind.

She bit the second pill in two and downed half of it. The other half she handed back to August.

"Gee, thanks," he said, but slipped it back into the vial.

Frida was ready for the high to slink upon her. It reminded of her being a teenager, when she'd nurse joints until the world felt different. Once she was rightly stoned, she'd go and make dessert. By high school, baking had become a kind of obsession. She'd plunge her hand into a bowl of silky, sifted flour, so high she thought she was communing with the stuff, and she couldn't wait to taste the cake at the end. She liked to bake all night, and at some point her mother would walk in and tell her to finish up, it was time to sleep. She often missed her morning classes, and her mother was too crazed to even notice.

"Do you always have drugs to trade?" Frida asked. Already, she felt the world going loose and dreamy.

August shook his head. "Rarely, and if I do, it's this playground stuff."

"I do feel like a kid again, even if this isn't weed."

She closed her eyes, opened them.

"Don't tell Cal, okay?" she said.

"Tell him what?" he asked, and she caught her reflection in his glasses. She looked drawn and tired. Jesus, what was she doing? Endangering the life of her child? *Oh, Frida,* she told herself. *Relax.*

They were standing on the other side of the carriage now. She put her palm on the mare. She felt a peace emanating from the center of her body. A mellow.

"I was much less tired looking when Cal and I met."

She wanted to complain about their stupid fight, tell someone, but from August's pause she could tell he wouldn't care to hear any of that. She'd wait him out.

"How did you two meet?" he finally asked.

His question was innocuous—maybe he wasn't really interested in the answer—but now she would have to talk about Micah.

"Through my brother," she said. "They were roommates in college." But Plank wasn't just any college, and that would need explaining, too.

"Your brother. Huh."

"What's wrong with that?"

"Older or younger?"

"He's dead." She hadn't meant to say it aloud, and that was probably obvious to August, who, for a moment, remained silent.

"I'm sorry," he said.

She didn't want to answer, but she knew she would, and that she'd tell him everything.

"He was a suicide bomber," she said.

"Shit."

She could tell he was truly surprised. Perhaps August had been a vagabond at the edge of civilization for so long that, for him, history was news.

It was true: Micah had strapped dynamite to his chest and blown himself up. He had killed thirty-one people and injured many others. Everything else about him was merely postscript, and the same probably went for Frida. She was the sister of a suicide bomber, the guy who blew himself up at the Hollywood and Highland mall, the man who had yelled, "Listen!" before pushing the button that set off the timer, which set off the explosion.

"He was the first to do it in L.A.," she said, "which made him . . . notable."

After Micah, she explained, people were killed at the supermarket, at all the other malls, at gas stations. Los Angeles, San

Francisco, New York, Boston, D.C., Chicago, Memphis. The good cities, all of them rendered violent and terrifying by men and women who martyred themselves in the name of—what?

"I thought I got over it," she said.

"But you didn't,"

"Right, I didn't. Not really. I put on a brave face, you know? For my parents."

"Understandable," he said.

"I haven't been able to stop thinking about him." She paused. "Cal brought him up this morning. That's why he's on my mind, I guess."

Maybe now she'd tell August about her possible pregnancy. He'd be especially nice, maybe offer her some free stuff. Or he might be mad she'd taken the Vicodin. Cal certainly would be.

The pills were doing their job. Her nose was tingling, and the space above her upper lip had started to itch. Her tongue felt a little thick, and so did the air. Things seemed so calm; it was as if the whole world had slowed.

She thought she could sense her parents, Hilda and Dada, nearby, as if they'd just gone into the house to get something.

After Micah died, Hilda wouldn't come out of her bedroom. She started to stink, and her hair hung in greasy strips around her sagging face. She kept refreshing the news pages, trolling for articles about Micah. She would leave anonymous comments, sometimes trashing her own son, calling him evil, sometimes celebrating what he'd done. She used all kinds of usernames, played all kinds of roles, became other people, told Frida it helped somehow. Dada didn't go into the room very often, and Frida had to take care of everything. Two years later, her father had called *her* the traitor, for planning to leave L.A., for planning to leave them. He wouldn't forgive her for abandoning the family. He had already forgiven Micah.

August cleared his throat, and Frida shook herself back to the present. She realized she hadn't said anything for some time.

"Sorry."

"I read somewhere about Iran, back in the day," August said. "They had these backpacks. For kids, you know? Decorated with pictures of suicide bombers like they were SpongeBob. Remember that old cartoon?"

Frida shook her head. "My brother was in the Group."

August squinted, like he was trying to figure out what that meant. Then he said, "Was he one of those pissed-off students?"

So August hadn't been gone so long; he knew what the Group was. According to Micah, the L.A. contingent had emerged a year after the earthquake, mostly college students who had been left with insurmountable debt and no way to pay it back. Nobody knew about them back then, or they did, but they didn't care. In the beginning, the Group was concerned that the city was still in shambles: collapsed houses and condemned schools everywhere, and the 101 severed in two at the 110. The Group couldn't believe the rich were complaining that their own neighborhoods weren't getting fixed fast enough, especially when it seemed like the only areas of the city that functioned at all were the affluent ones. A few of the founders were interested in politically motivated performance art; it was a means to get attention, they argued, a more interesting way to express their dissatisfaction. That was the theory, at least. Half a year later, the first Community opened, and people still hadn't heard of the Group. It had taken a long time for anyone to notice them.

A few months before Micah's death, Frida had convinced him to come over. She told him Cal would be at work. They'd gotten drunk on the bathtub gin he'd brought in his jacket pocket. Micah had insisted they enjoy it in the alley below their unit, and Frida complied because she never got to see him and she didn't want to scare him off. By then, her brother had become very particular about how he spent his time.

It was on that visit that Micah had told Frida the Group's origin story. She hadn't even asked; in her memory, he took a sip from his flask, leaned against the stucco wall of the neighboring apartment complex, and just started talking.

"A year after the quake," he said, "some wealthy douche bags took their stupid Range Rovers or whatever and surrounded an ambulance. They wouldn't let it south of Pico. Do you remember that?"

Frida didn't. She wondered if this was more legend than history, but of course she didn't say anything.

Micah went on: "Those assholes said whoever was dying hadn't paid for those services." He snorted and passed her the flask, which she took, grateful. "They didn't want to share, didn't care about anyone but themselves." He smiled then, his eyes glistening. "The Group was born right after that, to fight that kind selfishness, to keep people empowered." When Frida said nothing, he said, "Or at least amused."

"'Amused'?" Frida said. "But how?"

"A couple of the founders were into theater and performance theory, shit like that. Their early stunts were a little silly—I'll be the first to cop to that. Sure, maybe skipping around in a Mexican wrestling mask gets more attention than a regular old protest, but it's hard to be taken seriously when you act like that. But, then again, maybe that's what's kept us from being shut down." He paused. "Nobody saw us as a threat."

Micah reached for the flask, and once Frida handed it to him, he slipped it into his jacket and said, "Time to go." He pushed himself away from the wall, gave her a cursory hug, and left. Like that. She hadn't told Cal about the visit, but she was drunk and cagey when he got home, and he'd guessed.

The Vicodin was bringing this all back vividly. She shook her head at August. "Micah wasn't one of the Group's founders, but, yeah, you're thinking of the right people. He only got involved after he finished school. He didn't have debts like the first members, but I guess he had that same . . ." She searched for the right word. "Anger?" Frida paused. "By then, you know, things were a lot worse."

August's blank expression suggested that, no, he didn't know. "What about you?" he asked.

"Me? I wasn't in the Group."

A bird cried in the trees somewhere.

Usually, she was angry at her little brother for believing that strongly in the Group and its edicts: that money only poisoned people, that government was just bureaucracy, corruption, and oppression, that working wouldn't save them, only engagement would. Micah was always using that kind of language near the end—*engagement, engaged.* The Group didn't have a manifesto, or if it did she wouldn't know; it was so secretive by the end. But some of the members acted like a single unified truth was leading them forward. How could her brother value that stupid truth more than the blood pumping through his veins, his own beautiful, delicate joints, the intricate machinery of his breath?

Frida sighed and glanced at the bra she'd given August, now slung over the side of the carriage. "If bras are so in demand, does that mean there are a lot of women out here? How far do you travel? How many of us are there?"

"You know I won't reveal my route." He laughed. "Besides, that bra you gave me is made of fabric and wire, both valuable. And those little metal clasps, those annoying things? Also in demand."

"So it won't be a bra anymore? Is that what you're saying?"

August said he didn't know for sure how it would be used.

"What if I followed you?" she asked.

"Too dangerous."

She nodded. They had stayed put since their arrival here; the Millers had never ventured beyond the big lake four miles off. Everyone seemed afraid, and August had stoked that.

Normally, August was eager to pack up and leave, but she could tell by the way he lingered that he wanted to hear more about her brother.

"He'd been in love with an idea," she said. That was true, wasn't it? He wanted to rebuild L.A., neighborhood by neighborhood, and then he could do the same elsewhere. Not to make money, but to make the world livable again. Before he died, he'd helped set up the Group's encampment near Echo Park. Every month more and more people were leaving for the Communities, but poor disheveled

L.A. hung on. Micah thought the encampment was proof that people would be okay, even without money; anyone could come live there, he said, as long as they helped keep it running. He'd rebuilt housing, and the Group had spearheaded farming efforts in rambling, verdant Elysian Park; those fields needed crops, even Cal agreed on that. Rumor had it the Group would break into Dodger Stadium, too: the members would clean it up and make it a meeting place.

It all would have sounded idyllic if men with guns didn't patrol the perimeters, if they hadn't kicked out a bunch of residents who didn't want to cooperate, taken their houses and clothing and anything the people couldn't run away with on their backs. At least in Frida and Cal's dark and cold apartment a few miles west they could say whatever they wanted about anyone. Cal, who was still earning a meager sum to run a garden, said that all the zucchini in the world wouldn't lure him into their weird compound. "That encampment's almost as bad as a Community, except the amenities are worse," he'd said once. "Everyone there is giving up something. We just don't know yet what that something is."

The truth was, Frida wasn't sure why Micah had killed himself. All she knew was he had wanted to save the world, and shake it up. He was engaged. He was devoted to his cause, whatever it was. He didn't care about how his family would feel upon hearing the news. Even though she loved her brother, she couldn't help but think that only a monster could put aside the personal.

Some days she missed Micah so much, she didn't care what he'd done. She remembered how, in junior high, he'd made her a mask out of cardboard and tissues that looked like a bear vomiting streamers. How, for her fifteenth birthday, he'd sung "Happy Birthday" to her backward, *You to birthday happy*, his voice a terrible warble that had her laughing until she cried.

"I hope it didn't hurt," she said now. She could hear her voice dragging, the words slurring a little. "I hope it didn't hurt when he blew himself up."

She didn't tell August that she'd pictured his death too many times to count. Imagined the duct tape tugging at his T-shirt, the weight of the explosives on his stomach and chest. Why had he said, "Listen!" to the crowd of shoppers? Every day she invented a new sound, a different story, he wanted his victims to hear. Listen.

"It was probably quick," August said.

"You think?" Frida said.

Before the earthquake, when Dada was still working and Hilda didn't have to scrounge around for cleaning jobs to keep the family alive, Frida would smoke her joints and bake, and Micah would be in the living room, studying like it was breathing. He'd always been the smart one. No one contested it. Frida could have treated him like a pest, a nerd she'd rather disown than have to talk to, but it had never been like that. "Don't ask," she told her friends when they wondered aloud why she didn't just kick him off the couch so they could chill there. Sometimes they asked if she and Micah were twins, not because they looked alike or seemed close in age, but because they communicated like siblings who had shared a womb: wordlessly, without strife, accustomed to sharing the most limited of spaces.

Frida marveled now at how carefree she had been then. She didn't have any idea what would happen. The world was already going to shit, but it had been going to shit for countless generations before her. Overpopulation, pollution, drought, disease, oil, terrorism: all of that existed in the background, in the distance. Frida never read the news. She was fifteen and stoned, and it didn't matter if college wouldn't be there because she could bake her way to adulthood. She would run a shop or be the head pastry chef at a restaurant, or she would have her own cooking site. The future existed, especially for her.

"Once the bell is rung, you can't unring it," Micah had said that night they got drunk in the alley. *You can't go backward* was what he meant. But at fifteen she hadn't understood that—had she even understood it at twenty-four, when Micah died? She actually had thought geniuses were working to repair the world. Stupidity had protected her. The bell had been ringing all that time, and she hadn't heard it.

Sue brayed, and Frida wondered how long she'd been standing there, staring off into the distance.

"Maybe your brother felt good," August said. "When he died."

Frida should have wanted to cry, out of disgust or shock or sadness, but she felt nothing. Bless these drugs, she thought, bless this feeling.

This is what she'd always wanted. A painless life.

4

Cal returned at dusk with four pathetic chanterelles, which looked so much like a dead man's ears he didn't want to touch them, let alone eat them. According to Bo, foraging was women's work, and although Cal enjoyed the intensity of the process—the rooting around in soil, the animalistic obsession of it, his brain instructing him to *seek, seek, seek*—he thought maybe Bo was right. In the year and a half they'd known the Millers, Frida had become an expert at foraging. Under Sandy's tutelage, his wife could spot fungi and berries where Cal saw nothing but trees and brush, and though he would never admit it, he found joining her on a foraging trip frustrating. Nobody likes to feel useless. On their most recent expedition, they had returned home with her bag nearly bursting, his nearly empty, and she had said, "It's okay. You were my arm candy." He had scowled when she'd said it, but tonight that kind of teasing would be a relief. Even though there had been no more arguing, the tension remained, and Cal wasn't sure how to shrug it off.

The solar torches lit his path to the house. By sunset each day, one of them was supposed to bring a couple of torches inside, to illuminate the room. Without them, there would be nothing but blackness. Frida must have forgotten to do it. Don't worry, he told himself. She had not fainted or been kidnapped.

She had not been mauled by a bear or stung by some deadly mosquito. She was safe inside. She was just flaky. Always had been.

As he got closer, he saw that the front door was open to let in the last light and some air. And him, he supposed. Was that a sign that she'd forgiven him? Or that she had forgotten about him all together? The door was a mouth, and if he passed through it, he would fall into the dark throat of night. He shivered. Once the sun went down, he could easily imagine an evil out here. A stranger could come after them, a Pirate in search of food, tools, blood. Or a coyote might step through their open door, tongue out, eyes squinted. They weren't safe, not ever. He hated to think that way, but it was the truth.

He yanked two torches from the ground and made his way forward. He wished he had his gun, but he'd left it behind for Frida.

Cal had purposely stayed away for as long as possible, so as not to overhear Frida's conversation with August, her confession about the pregnancy. As if August were a priest, or even the pope. August was powerful: he knew everyone, could travel freely, and had probably heard everyone's secrets. But why did he have that privilege? That burden.

"Babe?" Cal called out. He said the word lightly but not obsequiously; he had apologized when Frida returned from the well this morning, and he wouldn't do it a second time. He was keen on getting past their little quarrel, and he hoped she was, too.

He tossed the mushroom bag on the card table in the kitchen area. He placed one of the torches next to the washbasin.

He heard Frida suck in her breath, not from their bed, but by the cots that Jane and Garrett used to sleep in. He shined the torch in that direction. She was on the floor, lying on her back with her hands behind her head, her legs twisted like a pretzel. Was she doing sit-ups?

"And then there was light," he said. He smiled. She was safe.

"I see that." She began bicycling her legs frantically.

"Are you okay?"

"No. I mean yes." She laughed; there was mischief in it, he thought. "I guess I'm just feeling antsy."

"Is it anxiety?"

She sat up, rubbed a hand across her face, as if to wipe something away.

He helped her to standing. Her eyes were pink. "You look terrible," he said.

"Thanks."

"Not like that. Sorry." There was the accidental second apology. He wanted, stupidly, to take it back. Instead, he laid the torch on the cot and, after a moment, took her in his arms. She let him. She felt fragile and hard, like a marionette.

"I got the garlic," Frida said. "For free."

"So you told him."

She didn't answer.

"How did he react?"

Frida sighed and leaned away from him. "Can we talk about this later?"

"Why?"

"Because it was so anticlimactic. He doesn't have tests, he doesn't know any midwives. It's not like he's tight with a shaman. And if he were, he'd never introduce us." She picked up the torch and leaned it against the wall. This one they would leave here as their night-light. It would fade before dawn. Without it, back when they lived in the shed, Cal had felt his very limbs disappear in the merciless darkness.

"He wasn't worried? Or excited?"

"He let me ramble," she said. "And now I feel embarrassed."

"Don't be," Cal said, but he could understand it. There was something about August that made you want to confess, and his silence kept you talking even after you wanted to shut up.

They began the nightly task of setting the card table for dinner. Frida peeked into the mushroom bag but said nothing disparaging about its contents or lack thereof. She seemed to move about the room as if in a fog, humming along to herself. In Cal's youth, his mother would sometimes stay up all night, editing a local commercial for extra cash or designing banner ads that no one ever clicked on, and in the morning she'd say, "I'm out of it," with a look that conveyed that he was, conversely, *in* it. That's what Frida was doing now, with her little floaty movements, her lack of conversation, those strange sit-ups. She was sending herself out of it. At least she didn't seem angry anymore.

"I want to know where August goes," she said suddenly. She held the bowl of beets aloft, like a trophy, and with an exhale placed it on the table. The torches gave the room a streetlight glow, a marry-me dimness, but Cal was used to it by now, sick of it even.

He sat down without responding. They hadn't had this discussion in a while, but Cal realized he'd been waiting for Frida to bring it up ever since she told him she'd missed her period.

"You know it's too dangerous," he said.

"So says August."

"You don't believe him anymore?"

"It's not that."

"Then what? We have no idea who's out there. What if there are Pirates? Don't tell me you aren't frightened of them."

"I've never run into a Pirate. Have you?"

"You know I haven't."

"So we still don't know if they're out there. But we know

someone is. And August, with his goods to trade, his vague answers. I'm sick of it."

"I want us to be safe, Frida. That's what matters most."

She said nothing.

"I know you're mad at me."

"Oh, stop it," she said.

She began to serve the beets, and Cal did the same with the sprouted beans. They were healthy, necessary to surviving out here, but they tasted terrible.

"I'm just curious," she said. "Don't you want to know what they do with the bras?"

"What bras?"

"Before," she said. "I traded him a new bra."

"Before when? And for what?"

"It's not important. What matters is what's out there. I need to know, Cal. Don't you?"

"If we leave, who will protect this place?"

She snorted. "Maybe we can find another house to steal."

Cal was about to bring a bite of the awful, humid-smelling beans to his mouth, but now he paused. "Don't do that, Frida," he said.

"Do what?" she asked. She was acting like a kid playing with her food.

"The Millers are dead," he said. "Get over it."

So this morning she was upset about Micah, and now she wanted to argue about their living situation. And August. She could be worse than a drunk, teasing for a brawl.

He put the fork in his mouth because if he didn't, he would say something nasty, even though his only true impulse was to protect her. Not that she wanted his protection; Frida never wanted any man's protection. To her, the whole idea of chivalry was pure self-congratulatory bullshit. She couldn't even abide the phrase women and children first. The night they met, she had asked, "You know why they don't say 'men and children first'?" He said he didn't. "Because that would be redundant," she replied.

Cal had let her get away with that one, but only because she was naked, and because they'd just slept together for the first time. It was commencement weekend at Plank, and Frida was a stowaway. Micah was getting smashed with the other graduating Plankers, their families already asleep in their motel beds two towns over. No one was there to see Cal graduate. His mother had died a few months earlier in the big snowstorm that decimated Cleveland, and his father, who lived an hour away from her on his little farm, had either disappeared or was dead, too. Cal tried not to think about it.

When he and Frida met, she asked after his family. "Who's coming to the ceremony?"

To his surprise, he told her the truth, and she said, "I'll be your guest." By then, he had heard so much about this Frida, Micah's famous older sister: the baker, the badass. And there she was: red lipstick like a glamorous wound, big white teeth, those sparkling eyes. Her strong, pointy chin. Those wide hips he wanted to kneel before, like a vassal. Falling in love with her had been easy. And, now, sitting across from her, eating this wretched meal, the same one they'd had for six nights in a row, he still loved her. He would take care of her, even if she didn't realize that was what he was doing.

Years ago, his father had hit a deer on the highway. He'd told Cal to stay in the car. "Turn up the music," he said. Cal did, and like a good boy, he kept his eyes on the air-conditioning vents as his father walked with a tire iron toward the suffering animal. "I put the deer out of its misery," his father explained later, and Cal felt grateful that he didn't have to watch or participate. His father had taken care of it.

He'd kept Frida from seeing the Millers' bodies because it was too horrible. He had taken care of it, but she was resentful. She acted like he'd killed the Millers, even though all four of them were dead in their beds when he found them, the covers drawn up to their chins as though they were waiting for Santa Claus, visions of sugar plums dancing in their heads. The rest of the poison sat on the kitchen table, white powder he didn't recognize in a glass bowl, waiting there like some sick invitation to death. He knew it was poison by its smell—the lack of one.

Don't drink the Kool-Aid, Cal had thought.

When they first began spending time with the Millers, Cal had occasionally worried that one day their tiny world would collapse: someone would be attacked by an animal or catch an illness that turned deadly, or maybe fifteen years down the road one of the kids would run away to the nearest city, or whatever was left of it. He didn't think it would happen like this, though. He had never once imagined suicide. For months afterward, Frida kept asking, "Where did they get the poison? Why not use something natural, like nightshade?" She wanted to know why Cal had gotten rid of the powder before she could see it. As if she didn't believe him.

He'd dragged their bodies out one by one and buried them. He remembered thinking how much Garrett had grown since the first time he and Frida had met him; the boy was now four and seemed tall for his age. He'd stay that way. The thought had made Cal sick. He'd tried to throw up, but he found he couldn't.

It took all day and half the night to bury them, and he had injured his back. He'd focused on the pain, imagined it as a thick red belt along his waist, because it was the only thing his mind could handle. He would keep it together.

He had returned to the shed the next morning. Frida was

crying, just about out of her mind. She thought something had happened to him. Something had. "We're moving," he said, as if he'd simply been house hunting.

They waited two weeks before they dragged their stuff to the Millers' place; even then Frida thought it was too soon, that it was disrespectful, greedy even. Cal told her it was wasteful to let the house sit empty like that, that Sandy and Bo must've wanted it this way.

"They sent me to find them," he said, but that was all.

He supposed he'd always withheld things from her; sometimes the whole story should not be repeated. He wouldn't describe how it felt, to carry those children.

Frida thought that the worse things got, the more women lost what they'd worked so hard to gain. No one cared about voting rights and equal pay because everyone was too busy lighting fires to stay warm and looking for food to stay alive. "It's like the only thing that matters anymore is upper-body strength," she complained. "Brute force." This was before the Millers died, when he told her he could lift the firewood on his own, warned her she'd get a hernia. Frida had been pissed, but, really, what did she have to be angry about? Yes, they had to rely on an antiquated division of labor. And yes, she would be rescued first from a sinking ship. Wasn't that a relief? "Spare me your white man's burden," Frida had said—which reminded him of Micah, in their Postcolonial Sexualities course.

Cal was sure it had been the worst for Bo. He had probably been the last to eat the poison. Someone had to make sure their dosages were correct and that his children, his wife, wouldn't awake. It must have been terrible. Who *wanted* to be a man?

Cal looked at Frida across the table. She was scraping at her food with the fork, her focus anywhere but on him. He cleared his throat. "Frida," he said.

"What?" she asked, and looked up.

"I have no interest in finding out what's beyond the territory we've already explored." He paused. "All I need, all I want, is right here. With you."

The only sound was Frida's fork hitting her plate.

"I hate these beets," she said.

Though he agreed, Cal shook his head. "You have to finish them," he said. Already the possibility of their unborn child was exerting its influence. It needed the nutrients.

———

Cal had last seen Bo alive six months ago, when the Millers had come to the shed for dinner. It was the beginning of spring, and Cal had again found himself missing the jacarandas in L.A., which had to be blooming soon. Two years earlier, they'd left town before the trees blossomed, and sometimes he imagined the purple flowers pastel against a cloudless sky. He used to love that, and how, come summer, the sticky flowers would carpet the sidewalks.

On that visit with the Millers, though, he felt at home. He didn't miss anything. The weather was warm, and the sun was a neon peach in a charcoal sky. He had roasted cauliflower, and Frida had steeped jugs of water with lemon balm. Thanks to the garden's bounty, they'd been getting crafty with their meals. They could have been back in L.A., throwing a dinner party. In the year and a half since they'd met the Millers, they'd learned a lot. They were getting the hang of things; that, or they had let themselves be fooled.

The Millers had brought a tent for sleeping, but Cal remembered waking the next morning to their absence; the family had risen before dawn to be on its way. They probably hadn't even gone to sleep, Cal thought. He imagined Sandy and Bo alternating security shifts, the chilly wind the only thing keeping their eyes open. He envied and derided their brand of dedication. Breathe out already, he wanted to say.

Still, the families were getting comfortable with each other. Garrett sat on Frida's lap occasionally, and the couples shared a few inside jokes. Bo and Cal had already gone hunting a handful of times, and Sandy had shown Frida how to forage for mushrooms and berries. She had explained to his wife how to distinguish between the poisonous and the safe.

That night, while Frida helped Sandy tuck the children into the tent, Cal took Bo up on his offer of moonshine. The liquor tasted like Windex, but he drank it anyway because he wanted to feel that old familiar ease in his brain and limbs.

It didn't take long for Cal to feel a little drunk. If he weren't, he wouldn't have asked Bo what was beyond their land. The Millers never spoke of this, even as they approached other topics: the local flora and fauna, visits from August, how to keep animals away from the garden. They were full of advice, and yet continuously evasive. *This place of mystery,* Cal reminded himself. But he wanted answers.

He could tell Bo was tipsy by the way he lay back on the faded quilt they'd been dining on and propped himself on his elbows. In the daylight, Bo had an alert and serious face, but in the darkness it was hidden, and he simply seemed small, and thus more vulnerable. Frida liked to remind Cal that their neighbor was short, as if this spoke of some deeper lack. "This isn't a nineteenth-century novel," Micah had liked to say back at Plank, and Cal thought of that now. Bo could just as well have had a wooden leg, he thought. It would mean nothing.

"Have you searched for others? Who the hell is out there?"

Bo didn't respond, and in the dark, Cal couldn't see whether he intended to. So he went on. "There have to be more of us. Why haven't they shown themselves?" He paused. "Why haven't they killed us and stolen our shit?" He and Frida asked themselves this all the time. Besides a lunatic who had jumped in front of their car on their way through the Central Valley, they hadn't been bothered by a soul since leaving L.A. Where were all the marauders—the Pirates—that everyone in L.A. was so frightened of? It couldn't just be luck that had kept Frida and Cal safe.

"Why do you assume they're bad people?" Bo asked.

Cal laughed. "I've seen movies."

"They trade with August, they keep to themselves."

"But why?" Cal asked. "What do you know?"

Cal heard the slosh of the liquor in Bo's Mason jar, and a strong gulp, as if the man were preparing for a long story. Cal waited.

"When Sandy and I first came out here—years ago—we went on an exploratory mission. We were still living in the shed, and we secured it with this big bicycle lock before we left. Not that it would really keep anyone out, but we figured it was more important to know the area, dangerous or not."

"Were there Pirates back then?"

Bo sighed. "We were curious, like you."

Cal wondered why Bo was being so shifty. He thought he could see that Bo had his eyes on the tent, where their wives were hushing the children. He seemed suddenly anxious that Sandy might hear them. Would she contest his version of events? Maybe she had another story to tell.

"We walked due west for days," Bo said, "and found no one, nothing. Nothing human, at least. Then we retraced our steps, and when we reached the shed, we went in the opposite direction."

Bo paused, and Cal forced himself to remain silent, to wait him out.

"On the second day," Bo began, as if he were reading from the Bible, "traveling east, we found a sign." He paused, as if this should mean something. As though this was a well-practiced script. Cal had no idea what he meant. Was it a simple octagonal stop sign or the Virgin Mary, burned into a rock?

"There were large spikes coming out of the ground."

"What do you mean, 'spikes'?" Cal imagined a line of them, like at the exit of a parking lot. Severe tire damage, he thought.

"They were huge," Bo said. As he and Sandy approached, they saw that the spikes weren't smooth and uniform. They were made up of cast-off objects—car parts, old clothing, plastic—and wrapped in barbed wire, their tips sharp and jagged. They were twice as tall as Bo, and they leaned, as if into a strong wind. "They were menacing. Their presence meant, *Turn around. Go away.*"

Bo and Sandy only wove their way around a few. There were a hundred of them, easily, but if they'd had the time—and the courage, Cal thought—they could have discerned a route through them. Not all of them were spaced closely together. If you knew how, you could get in and out.

"You know all those contested nuclear waste sites?" Bo asked.

Cal nodded. When he was a kid, there'd been endless debates about where to store radioactive waste. He remembered politicians winning votes by promising to fight the proposed projects—not that they could. The fear of another Chernobyl or Fukushima or Tarapur wasn't as strong as the need to put the radioactive material somewhere. Plank's campus hadn't been too far from a disputed site.

Cal took another sip of the liquor, and it burned down his throat. "What does this have to do with nuclear waste?"

Bo explained that experts in the previous century had designed different ways to warn of a site's danger, so that anyone might understand them: the foreigner, the illiterate, the alien. Large spikes had been one suggestion. In a thousand years, the message had to be clear, so that people understood what had been left there. "For the future," Bo said, and a thread of ice inched down Cal's spine. The future had arrived.

But the government had ultimately opted for something predictable; they'd plastered the sites with multiple signs bearing scientific information and stamped with the traditional nuclear symbol. Some said that future generations might take the image for an angel if they didn't know better. "Tough shit for them, I guess," Bo said.

"So these spikes you and Sandy saw? It wasn't a nuclear waste site?" Cal asked, confused.

Bo shook his head. "Doubt it—but they reminded me of one. As if they'd been made in homage to a rejected vision."

"So if the government didn't build them, who did?"

"I don't know," Bo said. "But they weren't that old. We found footprints, barely faded, and someone had dropped a leather belt, buckled very small. They must have been using it as a stirrup."

"So what did you do?"

Bo seemed surprised by this question, as if its answer were obvious. They went back to the shed, he said, and began building the house.

"Why did you build so far from the shed?"

Bo laughed. "We conceived of the shed as a hidden shelter, should we need secondary protection. Safer that way."

"Is there reason to be afraid?" Cal asked. "You still haven't told me much."

"I'm getting to that," he said.

The first time August approached in his chariot, Bo and Sandy assumed he was from the Spikes, as they'd begun to call them, and Bo stepped out of their house with his rifle. It wasn't exactly a house just yet, only its skeleton.

"August was here to trade. He wouldn't say much about where he was from, only that he was a middle man. He told me he liked to make sure everyone was getting along all right, that no one was sick." Bo paused. "He said the people who'd built the Spikes simply wanted to be left alone. They're separatists, and they don't allow their community to go beyond the border." The border, it seemed, was the Spikes; they'd built them long ago.

Bo told Cal he would have liked to get more information, but August was already off the buggy, showing Sandy his various goods. She wanted the shovel.

"It was like a shopping center for her," Bo said, and Cal thought, *Shopping center.* The phrasing had be a clue to where Bo was from; that kind of information was always lodged in speech.

Bo's story made Cal think immediately of the Communities. Gated, under surveillance, exclusive to everyone but the very rich. Back in L.A. Frida had often wondered aloud what they were like inside; before the Internet became too expensive and then stopped working altogether, she'd scoured it for information, for stupid gossipy facts. *No smoking allowed! All the houses look the same!* Some catered to Christians (mostly evangelicals), a few to Jews, while others didn't mention religion at all. All of them, though, claimed to have working electricity; clean, paved streets; excellent schools; and secure borders. If you lived behind those gates, the oil crisis was merely a nuisance. If you had money, you had everything.

Cal had just shrugged at Frida's interest, didn't want to

encourage her curiosity. Not like Micah, who loved to discuss them. The Communities made him murderously angry. They pissed off Cal, too, but he tried not to think of a world he couldn't enter.

"Did you ask August how many Spike People there are?"

"He wouldn't say."

After Cal's conversation with Bo, when the work of survival was backbreaking and difficult and the night a stinging kind of cold, Cal thought he and Frida might like being among the Spike People. Sometimes he felt the loneliness wrap around them like a net, especially once the Millers had died and they were living in their house. It was then that he wished to be allowed inside those spiky borders.

But he knew better. At the end of his story, Bo had leaned forward. "It's better to stay put, Calvin." His voice was stern, and then it turned ragged, almost desperate. "They're not afraid to use violence. That's what August told me. You stay put, Cal. You understand me?"

Cal said nothing. Frida and Sandy were headed toward them, their voices getting louder, closer. "Don't tell your wife about this," Bo whispered.

"Why not?"

"No need to worry her," Bo said, which had made sense at the time.

So Cal had never told Frida what he knew, maybe only because he had promised to keep quiet. He didn't want to scare her, but now she knew so little that she might do something rash. She didn't realize that they had to stay put in order to remain safe. Their curiosity would get them killed. How could he tell her that, without revealing all that he'd kept from her?

Cal realized now that Bo had known all along that he and his family would die soon. That's why it was easy to pass on the secret. Maybe it was a parting gift.

After the Millers had poisoned themselves, Frida and Cal spent a lot of time trying to understand their motives. But they kept coming up empty-handed. Had one of them been sick? Had they felt a sudden exhaustion with this life? Was someone after them? They only had questions.

By the time August found them living at the Millers' place, he didn't ask many questions. He had wanted to know where the others were, and when Cal said, "They ate poison," the man simply nodded and went on with his sales pitch, as if it were the most natural thing in the world.

"You don't look surprised," Cal had said to him, when Frida ran inside to grab something to trade. "About the Millers, I mean."

August merely raised an eyebrow.

"I'm not implying anything," Cal said quickly.

"I didn't think you were," August said. "But, no, I'm not surprised. Bo got the poison from me. He traded me his gun for it. He asked for assistance, and I gave it."

"Are you one of the Spike People?"

"What?" But then August understood. "Bo told you."

"Why don't they come here?"

"They believe in containment."

"Cut the cultspeak. I want to know."

It killed Cal not to have the full picture. How could he live in ignorance after he'd used every argument he had, every fact available to him, to convince Frida to leave L.A.? He'd told her there was a better world beyond than the one they knew. It was untouched; it had to be. A year before they left, another flu epidemic had hit the Northeast, and the population had been cut in half. (At least there was an upside of the oil crisis, people said;

73

disease couldn't afford to travel very far anymore.) The storm that killed his parents in Ohio had been followed by bigger and worse ones, and before the Internet went dead entirely, Cal read that only a third of the population in the Midwest and the South remained. "Anyone who's left is staying put," he told her. That was true in L.A., where people hung on to what was familiar. The city was rotting, it couldn't be denied, but at least it was *their* city. And even if people wanted to leave, the state of the roads and the rising price of gasoline kept most from doing it. Soon, the oil would run out. *Kaput.*

"What about the Pirates?" Frida had asked, many times. There were stories about people who had tried to leave town only to be murdered as soon as they crossed the city limits. Rumor had it that Pirates collected victims' teeth and hair and recycled them into household goods. Women were raped, people said. Men tortured. Cal didn't know what to tell Frida, except that there was no proof that the Pirates really existed. He'd researched it, asked around, and came up with only more gossip, more fear. First he told her they had to be a myth. Then he promised her they'd drive fast and that they'd only stop to refuel once they were safely in the woods.

"And I have the gun," he'd said. Someone who worked with Frida had sold it to them a few months prior.

On their way, Cal and Frida had been vigilant, but there had been no trouble. A miracle, Frida said at the time. They'd seen no one but that one harmless man, and Cal's theory turned out to be right: everyone left was either hibernating in the cities, waiting out hard times as if they'd ever end, or they were safe in the Communities. Or they lived out in the middle of nowhere and didn't want to cause trouble. Could that be the whole story, though?

"It's safe if you mind your own business," August had said

suddenly to Cal. "Don't kid yourself—they can't be bothered with you."

Before Cal could speak, Frida had returned, and August was back to hawking his wares as if nothing had happened in her absence. Cal would not ask August any more questions.

———

All these months later, he'd pretended he wasn't curious about August and the territory the man canvassed. Cal had hoped Frida would follow his lead, keep her head down, and focus on survival, on being happy in whatever way they could. Didn't she understand that safety was most important? Especially now, if there was going to be a baby.

He looked once more at his wife sitting across the table from him. Her plate was still full.

"Eat," he said.

She didn't reply, but she took a bite. Relief spun through him like a cure. She'd listened to him for once.

Cal knew it was settled: he would protect their family, whatever it took. He wouldn't say a word to Frida about what August and Bo had told him. He couldn't.

5

At dawn, Frida slipped out of the house before Cal could stir. She was headed to the creek to do laundry, a chore that now felt like a hobby.

By the time they'd gone to bed the night before, she no longer felt the Vicodin, but she still couldn't recall a single fragment of dream. This morning it was like her entire nervous system was wrapped in layers of gauze. She felt empty. At fifteen, she'd smoke until she hallucinated, and the next day, she would awake sharp as fangs. Now she was older, and her body had grown too used to being sober. It couldn't handle having fun, not like it used to.

She still enjoyed the walk to the creek: it was hard to be afraid this early in the day, when the world felt so new. In the woods, there were many mysteries: the cat's cradle of trees, for instance, why some fell sideways, as if pushed, branches lodged in the bramble, and why some lost their leaves and turned black, as if dipped in shoe polish, even as the others remained perfectly healthy. The pine needles still made her think of shredded wheat, though she hadn't had a bowl of that in almost twenty years. She loved the hushed quality of her steps along the path— Cal was religious about keeping it clear—and the sounds of the earth groaning. Even the rustling of small animals didn't bother her. If she listened closely, she could make out all the different kinds of birdsong: the beseeching, the joyful, the forlorn.

She passed a patch of mushrooms—how could Cal have missed those?—and turned right at the big redwood. The creek was down the incline. She could hear it now.

She hadn't spoken to Cal since yesterday's dinner, when he made his stupid pronouncement that all he needed was her, that they would not go exploring. She hated that he made these decisions without her, as if they were his alone to make. And why had he made this choice? She knew he was hiding something.

If they spent some time apart, if Cal spent the day digging and she did all the chores, each might find forgiveness for the other, at least start talking again. She just hoped their land wouldn't be riddled with dozens of holes by the time she returned. She didn't want to step into one. But did Cal? It felt that way, the way he'd acted.

Frida wasn't stupid. It was obvious he knew something, and that something was nailing him to their house, this tiny four-mile area. He was pretending to be content, but that was impossible. No one could eat sprouted beans in a dark house for days on end without complaint, without hatching an escape plan. They'd slept in Bo and Sandy's bed for almost half a year now, and every day the mystery of their deaths deepened.

August must have told him something. Or Bo had. While she and Sandy had been discussing their menstrual cycles, or the best techniques for mushroom foraging—Pussy stuff, Frida thought wryly—the men must have mapped out the territories, whispered state secrets. Regardless of who gave Cal the information, he wasn't sharing it with her.

Last night, once August had left, she'd gotten to thinking again about the outside world. Even after all she'd told him about Micah, what her brain kept returning to was the bra.

They, whoever they were, would cut it open and use its parts. The butchery of necessity. She imagined women with pendulous, aching breasts, and their children with braces built of Maidenform wires and clasps. Everyone in the tribe would know how to rehab bras—a command from the king. How silly, she thought. They probably made weapons. They were probably geniuses.

She must have seemed like such a moron to August, getting so high she wept for her dead brother. As if August cared about her stupid family drama. Oh goody: one more whiny white girl! Boohoo, Frida. But he had listened, hadn't he?

The creek would be stunning at this hour, and she moved faster to reach it. Sandy was the one to tell her that the morning was the best time to wash clothes, because it was cooler. That way, they would stay wet until she got home to hang them up. "No use waiting around here for your panties to dry," she'd said, nodding at the water. But Sandy had never mentioned the dew on the grass surrounding the creek or the occasional deer, prancing carefree, or the coolness of the rocks at the water's edge, which hadn't yet had a chance to absorb the heat from the sun. Of course she wouldn't have; Sandy was a practical woman. This morning, a bunch of dandelions gone to seed sprouted from the patch of grass, and for once, Frida was glad Sandy wasn't around. She put down her bag of clothes, and plucked a fluffy white flower from the ground. *Make a wish,* she imagined her mother saying, and she closed her eyes. If Sandy were here, she wouldn't approve, and if Jane were with them, Sandy might grab Frida's fist and say, *Please don't.*

At least, when you were on your own, you made the rules. Frida closed her eyes and wished she knew more, then blew the fluff off the flower.

After high school, she used to scour the sidewalks for

dandelions. That was when she was helping to support the family by working at Canter's. She ran the deli's ancient cash register, and on her walks home she'd seek out a dandelion so that she could make a wish. She'd recently charmed her way into the kitchen, and the head baker was letting Frida shadow her, teaching Frida how to make bagels and rugelach and that banana-chocolate cake with the word banana in yellow icing across the top. Ingredients were getting expensive and hard to keep in stock, and so she was also taught recipe substitutions, how to make a little go a long way. These lessons happened after Frida's shift and didn't pay, of course. Most of the time Frida was wishing on her dandelions for a permanent reprieve from her insipid register duties, a way into that kitchen for good, so that she could start a proper grown-up life. Not that she was in a hurry; things were tolerable at home, and it wasn't like anyone her age was leaving the nest.

But most of the people her age weren't like Micah, who was smart. Brilliant. A kid who needed to get out of L.A. so he could return to save it. The whole Ellis family, not just Hilda and Dada, but Frida, too, expected Micah to solve this mess the world was in. Her mother had met a Plank alum at a party. "A man so shy he was rude," she said. But he was also very smart and successful. "He's working to fix the water crisis," she said. "You know, how to make sure we still have some in the years to come." Their mother thought Micah could become a water man and solve the city's problems. Plus, the college was free, and not so far away.

This was what Dada liked, that it didn't cost anything to attend. There would be no other option for someone like Micah, who couldn't afford the private colleges. *Scholarship* was an endangered word. Back then, some people were still getting into college, but fewer and fewer were going; it was becoming a path reserved for the very rich. When Frida had started her senior year, she told her parents she didn't want to go to college, and they were relieved. Why bother with all that schooling if there wasn't a job waiting for you when you finished? "You don't need college," Dada had said, which seemed like a compliment at the time.

Her brother, though, he was different. He *did* need it. By then, UCLA and Berkeley had shut down, as had all the other public universities worthy of Micah's attendance. "Budget problems: the understatement of the decade," he liked to say. He was fucking brilliant, and he'd been born at the wrong time.

It took a single Internet search to learn about Plank's all-male student body. Frida was convinced her mother had kept this fact out of the story because it would scare Micah off. Girls liked Micah, and he liked them right back.

But then one night at dinner, her brother came to the table, his face held solemnly, and said, "I'm applying." That was all. Frida realized then that his beard, which he'd just started to grow, and the books on homesteading and animal husbandry that he'd recently downloaded, were making him into the kind of man Plank would accept. That morning, he'd quoted Thoreau, and she hadn't thought anything of it. But he was preparing. He was cunning, her brother.

It wasn't until Micah moved to Plank that she realized he'd gotten away. She was still in L.A., still living with their parents. Meanwhile, her brother had gone to live a grown-up life.

Cal liked to describe Micah as a prankster at Plank, but Frida didn't see him that way. She would say he took on dares, and with that bravery he defied you to take on your own. Plank was a dare. He would become a person who could live without women, who could work a farm, who could live in the past. "And you will give up that stupid deli job," he told Frida, "or you'll ask them to hire you in a different capacity. No more validating parking tickets at the register, for fuck's sake." Frida waited until Micah left for Plank, and then she took his advice. She asked to be put in the bakery, or she'd quit. To her surprise, they promoted her right away.

Frida smiled now at that tiny coup and from her bag brought out the laundry soap. The Millers had left it. She had a feeling its ingredients had come from August, but until they ran out, she wouldn't ask. Frida actually looked forward to making detergent herself; she thought it might remind her of baking: the measuring and mixing. It made her heart ache a little. She had been so good at her job.

She remembered writing to Micah about her promotion. She had sent him a letter, because Plank didn't allow email. In reply to her news he had said, "I knew you could do it, Freed. When I'm home this summer, can you give me a few lessons? Our head bread maker is graduating in June, and Cal says the position should be mine."

That's how she learned about Cal, through Micah's anecdotes. In the beginning, she was jealous of this new roommate who seemed to take her place as Micah's main confidant, recipient of his advice, and sounding board for all of his plans, both ridiculous and ingenious. But soon, she began to look forward to the Cal stories, as if he were a character in a soap opera she followed loyally. Cal had stayed up every night for a week, reading. Cal was so clean, Micah thought he might be psychotic. Cal knew how to fix a fence, "like a goddamned rancher," Micah wrote. Frida could tell that her brother admired his roommate, which was strange, since Micah rarely admired anyone. That wasn't how Cal saw it; he told Frida that Micah had underestimated him from day one. Perhaps that's how things ended between them; but it wasn't how they began.

The sky was turning from the purple of dawn to a dazzling

blue. It would be warmer today than yesterday. Frida sighed; she should be drinking more water. If she were pregnant, she'd need to stay hydrated, and either way, she didn't want a headache. Thank goodness the water around here was clean—or at least clean enough. She dipped her cupped hands to the creek and pulled the cold water to her lips. Out here, she often found herself dreading even the smallest physical discomfort. And now that she knew August could get her pills . . . well, that was just too dangerous. He'd told her such trades were a rarity. Was that true?

Frida brought her hand to her stomach. Did she want to be pregnant? She couldn't keep a child out of danger. But she'd love him.

When she and Cal had first started dating, he'd told her, "The only reason to bring a kid into this godforsaken world is to give it a mother." His own mom had died a few months earlier in that first awful snowstorm. It was a crazy thing for him to say, but Frida had loved it, had loved him, for being so mixed up. Cal. For a while after he moved to L.A., Frida couldn't abide anyone but him. He wrote her poems and brought her vegetables grown in one of the community gardens he oversaw, and at night they made frenzied love on her narrow bed, sometimes rolling onto the floor because there just wasn't enough space for this thing they needed from each other. If someone had told her then that the two of them would marry and come to the middle of nowhere to be alone—well, she would have smiled.

Like Jane and Garrett, their child would have no idea of the world he was missing. He'd think this, wringing out shirts in a babbling creek, was the height of entertainment. Her kid would grow hooves for feet like Garrett and Jane had, run through the woods with his eyes closed, and eat squirrel meat. The stories of Cal's mother and father, the artist and the farmer, would be myth. Hilda and Dada, just a fairy tale. So would, too, the terrible things they'd left behind to come here. Sandy and Bo had tried to create a new world for their kids, but it had been flawed. It was nothing compared with what Cal and Frida would build.

She sank a pair of her leggings under the water and then rubbed soap onto the waistline, let it foam up. The creek was so cold it made the joints in her hands ache. She'd have to pull them out soon and rub them in the dirt, or else her fingers would get too numb to work. A small fish flitted across her wrist, and her heart sped up, pulling her out of her brain fog.

She had a sudden desire to go running. She should be in better shape if she was going to give birth. She spent hours doing manual labor but nothing that really worked her heart and built up her body's stamina. She missed that.

In L.A., before it got too dangerous, and before the streets fell too badly into disrepair, Frida used to go jogging. On the first few attempts, her lungs had felt swampy, her breath at once sharp and shaky, and she had to stop every few feet to recover. It hurt. But she kept at it, and each time, she ran a little farther. Two weeks in, her body began to crave those miles.

She used to go with Toni, Micah's girlfriend, who was also in the Group. Soon after the two women met, Toni enlisted Frida to join her on her runs. "It's a great stress reliever," she'd said, and added that she didn't want to run with anyone in the Group because she was trying to deal with her "Jealousy Problem."

"Sounds like a bad movie," Frida had said.

"A juicy story, of love and loyalty. It would be quite good, actually," Toni said.

Frida laughed, but she knew Toni had truly been hurt by the way the other girls fawned over Micah, and how Micah lapped it up like a kitten before a bowl of milk.

"It's too bad I believe in the cause," Toni said. "Otherwise, I'd just leave him."

Frida remembered Toni's remarkable ability to run and talk simultaneously. She was short and muscular, a woman who might have had a childhood in gymnastics if she hadn't had such fucked-up parents and a grandmother who worried too much to let her do anything extracurricular. Toni wasn't even sure where her parents lived. Maybe they were in a cult in Boulder or off somewhere in Mexico. She'd been raised by her Nana outside Seattle. Toni loved her grandmother, but it was obvious she resented her for being so strict, instead of placing the blame on her parents, where it belonged. At seventeen she had run away from home. Micah was convinced that Toni's grandmother lived in a Community now; she'd told him her Nana had money. "Face it, Toni," Micah liked to say, "you're tony." Toni didn't appreciate Micah's joke and what it implied: that she could return to that life of comfort and denial at any moment. That she might. That she was merely slumming.

Before she fell into a friendship with Toni, Frida didn't know all that much about the Group; her brother would occasionally divulge a tidbit here and there, but that was it. Like everyone else, Frida remembered the Group for being responsible for a few pranks, which they posted online immediately afterward or streamed live. The Group may have been founded by disgruntled students who wanted to rid the world of corruption, but that didn't mean they knew how to get the public on their side; in those first few years, it was only the playful element that became visible to outsiders, making the Group's organized outrage hard to interpret. "Think of it as two different branches of the same tree" was how Toni put it. "The drama club dorks on one, and the more socially minded theorists on the other. How we ended up on the same tree is kind of puzzling. But then again, both branches want to disturb the status quo, make people pay attention. It's just a question of how to do that." She paused.

"The performance art folks were helpful in getting our name out there, and they have some ideas we can still use, but they're so naïve, not to mention unfocused. The truth is, the Group is growing up, getting more serious." She paused again. "To use the tree metaphor again: the artsy branch will eventually break off."

When Micah joined, the world had still only seen the pranks, the playful stuff. Dada called the Group an avant-garde theater troupe, and, at first, that was kind of true. They were famous for getting a thousand bicyclists to merge onto the 405 at rush hour. That had really fucked with whatever traffic was still left on that ruin of a freeway, but only for an hour or so. A pocket of the Group was made up of dancers, actors, and artists, and they'd done a few big performances in the middle of open trials and city council meetings.

Right after he graduated from Plank, Micah told Frida that he was moving into a loft with other members. "The Group?" she'd repeated, and asked if he'd also gotten into acrobatics and fire-breathing when she wasn't looking.

"You've got it all wrong," he said.

"It just doesn't seem like your thing," she said.

Micah had shaken his head. "You don't know anything about me."

That had stung a little, and still did. If she knew anyone, it was her little brother.

When had that stopped being true?

Frida pulled her hands out of the creek water, and the cold iced up her fingers. She crab-walked to a patch of dirt and placed her palms flat on the ground until the groan of cold subsided.

The other, more serious branch of the Group had always been there, but it wasn't until after Micah joined that it began to grow stronger. Or at least that's when she noticed the shift,

85

maybe because she started paying attention to their activities. Not long after Micah graduated from Plank, a few members of the Group had donned ski masks and hijacked a political fund-raising dinner. Those in attendance were said to be members of a nearby Community who wanted to close the roads surrounding their newly built compound. Someone from the Group ran a knife across a woman's cheek, scarring her, and another had bashed a man's head into one of the fake-orchid centerpieces. The Group had been protesting "corporate sponsorship of candidates," according to the signs they showed to the camera. When Frida asked Micah about how it related to the bike prank, or to the juggling of doll heads, he shrugged and said he didn't know a thing.

To Cal, it made sense that the Group appealed to Micah. "He's interested in social justice, or so he says," Cal remarked when she brought it up with him. "And he can also be dramatic, you know how he loves elaborate pranks."

Once she and Toni had been running together for a few weeks, Frida got up the courage to ask her the same questions.

"That's exactly what we discuss at our meetings," Toni said. "Are we undermining ourselves with our funny stunts? Or are we working toward the same goal?"

"And what exactly is that goal?" Frida had asked. They were running faster now, and she could hardly get out the words.

"Total world domination, of course," Toni said, and laughed. Then she said, "If people think we're just a bunch of clowns, we can get away with a lot more. Why do you think those morons let us into that fund-raiser to begin with? They must've expected a fucking flash mob."

Micah was never in any of the Group's filmed stunts—playful or otherwise. He claimed he wasn't holding the camera, either. Frida had watched the clip of the fund-raiser stunt over and over, despite how hard it was to do so, just to make sure her brother wasn't one of the masked offenders. He wasn't; she was sure of it. Besides, he'd never lie to her.

At that, Frida imagined the creek laughing at her. *You naïve little idiot,* it might say.

Frida dug her nails into the dirt—it felt strangely satisfying. It was a bad habit, and because of it, her nails were always filthy after doing the laundry. She stood up and returned to the creek's edge. She held her breath as she pushed a dress under the cold water.

From the beginning, Frida had liked Toni, who kept her hair in a tight ponytail and wore weird shoes like a revolutionary war general's: square and buckled. The night she met her, Toni and Micah had come over for tea made with mint from one of Cal's gardens. Her brother had barely touched his mug when he told Frida to stop watching the fund-raiser video. It was months old by then. "I told you, I'm not in it," he said. "I'm not an actor, nor am I a director."

"But you *are* a ham," Cal said.

"That's true!" Toni had cried, which made Frida laugh. Her brother needed a woman to put him in his place.

"I don't get what you're after," Cal said. "That poor fund-raising volunteer has a scar on her face. They said it got infected while healing. You know how hard it is to get antibiotics nowadays, and half the time they don't even work."

"The point is," Micah said, "people are waking from their numb slumber."

"It won't be long until we do more," Toni said, and Micah shot her a look.

"What does that mean?" Cal had asked. Frida remembered he suddenly looked very serious in their candlelit living room. They were sitting on big pillows on the floor, and the large chessboard they used as a table was between them, its brown and beige squares splattered with old wine.

Two weeks later, one of the gubernatorial candidates was kidnapped. After sixteen days, he was let go, naked except for a paper party hat, at the gate of the Community in Calabasas where he had thrown some rallies. He was unharmed, his campaign people said, but that could not be verified.

The Calabasas Community wasn't its own city, not yet, but it had exploited a loophole: it ran its own schools, funded its own police force and firefighters, and anyone hired to protect and work within its borders either had to be related by blood to one of its residents or pass a rigorous application process. But nobody knew how to apply because the details weren't on its website. Calabasas was apparently pouring money into alternative energies; it'd be the first carbon-neutral and energy-independent Community in California, which would make it even more attractive to prospective residents who were sick of blackouts and high energy bills.

The politicians understood that these were the constituents who mattered. Hardly anyone outside the Communities voted anymore. It didn't seem to make a difference. Some people were waiting for the Communities to become their own sovereign states. It was only a matter of time, people said. Micah hated to hear this. It wasn't a foregone conclusion, he said. They could fight it.

Cal had simply thrown himself into his gardening projects. He argued that if the rich forsook them, the country might be better off. "Maybe I'll run for office," he said, holding up a basket of onions. "I'll run on a vegetable platform."

He was joking, but Frida thought what Cal was doing made sense. He taught people how to grow their own food. This was necessary. After all, his expertise had kept them alive.

Cal would never let them go hungry, Frida thought now. He'd gotten them this far.

She lifted the dress out of the creek, and was surprised by how heavy the water had made it. Laying the dress across a rock, she grabbed Cal's pants, faded and dirty at the knees and still cuffed at the hems. Such a sweet sight, his clothing, wrinkled and wet, removed from his body. Even when things got difficult between them, doing Cal's laundry made Frida feel a love so tender she could weep.

The Group never took responsibility for that first kidnapping. It was obvious they were behind the stunt, though, and for an entire month Frida and her family didn't hear from Micah. Her parents had no idea what was going on, but they were too busy struggling with Dada's diminishing career and the cost of living to worry too much about him. Besides, he'd never been great at keeping in touch with them; and when he did swing by, he'd bring liquor and a crate of potatoes, and they'd be delighted, tripping over themselves with gratitude.

"They probably think he's doing summer stock," Cal joked.

Four weeks into Micah's disappearance, Frida had walked to the east side for answers. She didn't dare drive; the lines at the gas station were long, and it would've taken a week's worth of wages to pay for the trip. Besides, she didn't want Cal to know what she was up to. To this day, she'd kept it a secret.

She had turned onto Echo Park and walked a block when a man had approached her, empty-handed but imposing. "Can I help you?" he'd asked.

"I'm a friend of Toni's," she said, and the man looked at her closely before nodding. He whistled once, loudly, and suddenly there was her brother's girlfriend, calling from the window above.

Toni lived on the second floor of a ramshackle duplex that

overlooked Echo Park's now-drained lake. The lake's old bridge was gone, maybe burned for firewood, as were the pedal boats. Frida had been born too late to see the lotus flowers, which had once floated across the water's surface.

"Where is he?" Frida had asked Toni as soon as they were face-to-face. "Is he okay?"

Of course Toni wouldn't tell her anything, at least not there, not with other Group members in the living room behind her and hanging around on the porch below.

Looking back, Frida realized the Group had established a nascent encampment, even then. Everyone on that block was a member of the Group. That guy who had whistled for Toni was protecting their space. Already they were patrolling that part of town. Already they'd put people to work to improve their surroundings. Frida had nodded to the women in the empty lake who were picking up debris with yellow dishwashing gloves on their hands. She wondered if they were the same women who flirted with her brother. Toni had only smiled at the view. "We're all about beautification." Then she said Frida should go.

The next day, Micah showed up at Frida and Cal's door with a party blower in his mouth. He blew into the mouthpiece, and the striped plastic unfurled into a straight line with a *crunch* that made Frida's stomach tighten.

"You aren't dead," she said, and let him inside.

She did not give him the satisfaction of asking about the kidnapping.

That had been a rough time, Frida thought now, not even counting the collapsing economy, the nights without power or heat. Thank God for the weather in L.A. and the tiny apartment she and Cal had moved into; they kept each other warm. It was rough because of Micah's secretive life, and her parents' ignorance—their denial—of it; because of Cal's disdainful remarks about her brother, whom she felt a compulsive need to defend, and because of Toni and Micah's arguments: the damage of those fights trailed them like a pack of hungry dogs. And then Canter's closed, and Frida couldn't bake anymore. And then she and Cal had even less money. The day she brought home loaves of stale bread for the last time, her hairnet balled in her back pocket like some useless currency, she'd thought it couldn't get much worse.

But it could, and it did.

Frida pulled Cal's pants out of the water. Without thinking, she stuffed them into the laundry bag. It was stupid—she still needed to do the socks—but she suddenly wanted to be back home. A dark puddle spread across the bag, as if it had been wounded.

She didn't want to fight with Cal anymore.

Hilda used to say that anger was a choice. Frida could make the choice not to be angry with her husband, even if he was keeping secrets. She'd lied to him about August, hadn't she? Cal thought August knew about the pregnancy, but he didn't. Now he knew about Micah, and she had promised Cal she'd never tell anyone out here about her brother. She had broken that promise, and so easily. It made her sick.

She didn't want conflict to eat them from the inside out, as it had done to Micah and Toni. After a while, Cal had refused to spend time with them as a couple, so painful was it to watch them avoid each other's gazes, to use the other for sport.

Micah was the one to diagnose Toni with her "Jealousy Problem." It might have been a problem, but it didn't mean her feelings were unfounded. Toni had described to Frida what it was like to watch him with the other women in the Group. She didn't like the way they brought him little treats; once a girl named Leanne had stolen a bag of Jordan almonds for him. "For his long hours," Toni said, rolling her eyes. He'd eaten them in their bed as Toni tried to sleep. "I prayed he'd crack a molar." She didn't like the way the girls were so eager to volunteer their time for him: they'd gladly transcribe, email, or post links to go viral. And at the meetings, she said, it had gotten too much to bear. "They come earlier to get a seat in the inner circle," she said. Apparently, metal folding chairs were set up in concentric circles like tree rings with the meeting leaders standing in the center.

The Group met in an abandoned Taco Bell. They were squatting in it, and they'd move to a new space soon enough. Toni wouldn't say where the restaurant was, though how many could there be in Echo Park? Cal thought they were getting more careful, and from what Toni said, it sounded like it. Frida wondered when the other members would ask Toni to stop running with outsiders.

The Group had removed the restaurant's bolted-in booths and tables for their meeting space; only a select few ever saw what was behind the counter in the defunct kitchen. "They're building things in there," Toni said once, at the very end of a run, and Frida noted that she'd moved from *we* to *they*. Frida hoped to get more information from her, but soon Toni was back to talking about the girls. She was tired of watching the proceedings from the Siberia of the outer circle, she said, and she refused to arrive early. "Whenever Micah takes the floor, the younger girls, especially the new ones, lean forward, as if they're having trouble hearing him."

Micah had grown so secretive about the Group that he would have killed Toni for telling Frida all this. Cal either wasn't interested or he was derisive, and even if Frida had wanted to join the Group, Micah would have denied her entry. She had fallen in love with a man who had dismissed her brother's passion, and for that, Micah withheld everything from her.

One day Toni showed up at Frida's place and said she was done with the meetings.

"What happened?" Frida asked.

Toni said she'd stood to suggest an all-female delegation. She thought that might help things; let the older members tell the younger women how the system worked. She tried to present it as a way to strengthen bonds. No one seemed interested. Before Toni could even sit down, Leanne stood to ask Micah if he wanted her to mop the floors after the meeting.

Toni said she didn't mean to plunge her sharp, rusty bobby pin into the bitch's arm, it just happened.

After that, Toni was still a member, still involved, just not on a weekly basis. Or so Frida assumed, until the Group blew up the entrance to the county hospital. No one could prove they'd done it, but it was obvious to Frida it had been their work. The hospital had begun to charge for entrance into the emergency room—cash or gold only—and a man had died of dehydration in the parking lot. The Group had allegedly thrown a Molotov cocktail through the sliding glass doors. The man taking money at the entrance had been killed, and a nurse lost her hand.

Two days later, during a run, Frida asked Toni about their last stunt. "What the hell, you guys?"

Toni sped up. "I'm over it. The Group is Micah's thing." She didn't want to talk about it. She was no longer a member. Just like that.

Micah also wouldn't explain. "We've moved in a new direction" was all he'd say.

Not long after, Toni was supposed to meet her for a run, but when Frida opened the door, Micah was there. He'd shaved his head, and beneath a sharp stubble of hair, his scalp stunned white.

"Where's Toni?" Frida asked.

"She's gone," he said. He was very calm.

"Where did she go?"

"She left."

Frida thought he meant she'd simply moved out, but, no, she was *gone*. Frida never saw or heard from her again. Everyone guessed she had gone back to her grandmother in Washington after all. Sometimes, even now, Frida thought of Toni in one of the Communities. She was running down a smooth, paved road. There were cameras on every corner and uniformed guards, and she felt safe and clean. She probably had a baby.

Frida pulled the pants out of the bag again and shook them so they hung straight. She walked over to the big rock and lay them flat across its surface. It wasn't time to go yet. She wasn't finished.

Cal wanted Frida to be pregnant. And Frida wanted that, too, if she was honest with herself. It felt like a dare, the biggest, most important risk of all. Micah would think so. Before the Group ruined him, when his mind was still open, fluorescent as plankton, he might've written her a letter that said, *Go ahead, believe in it. Don't get all afraid on me.*

"I'm not afraid," she said aloud.

The way her voice sounded in the morning air made her turn around. Was it some desire for a reply? No one was there, just the trees. This didn't surprise her, but she did feel disappointed, as if she'd been stood up. But by whom? The creek rushed along, oblivious, and across the water, the forest waited. She and Sandy used to go foraging there, but they always stopped before getting in too deep, before the land became alien.

In a moment, she had the mesh bag full of Cal's socks, and she was crossing the creek. The trees seemed to step aside, to let her into the darkness. *After you,* they whispered. She walked farther, in a direction she'd never gone. But there was a path here, slightly overgrown, and she saw the track marks of August's carriage. She knew he carried a scythe to cut away brush as he traveled; it was as if he had cleared the way for her.

She draped a single sock on the branch of a tree. The fabric was gray and thin, and it had once belonged to Bo. Now it was a crumb that would lead her back. Hilda and Dada and Micah would be a fairy tale to her baby, but for Frida, this world here, the afterlife, was the fairy tale. If she wasn't careful, Frida would be eaten by a witch at the end of her journey.

In a few moments, August's tracks led to a narrow trail, thick redwoods on either side. Frida paused, hands clenching the bag of socks. She sighed. "I'm not afraid," she said again, as if to remind herself. She placed a red sock on a branch and kept walking.

Every few minutes or so, she left a piece of clothing for herself to find on her way back. And she *would* find her way back. The longer she walked, the more her chest tightened. She'd felt like this before, driving lost in L.A., her Navigator *and* her Device dead. She'd pass through a rough neighborhood hoping to find something familiar so she could breathe again, blink again, though, by the end, every neighborhood that wasn't a Community was rough. She was alert in that same way now. She had to pay attention, or she might get turned around, never find the thread of the route. The clothing wouldn't be any help if she headed in the wrong direction. She had created a system: colored clothes meant turn right, black and white ones meant turn left, and gray, head straight. She kept her eyes on landmarks: The tiny stream. The vines choking a thick trunk. A lone crocus.

After she had walked for about an hour, she saw something white up ahead. She quickened her pace, even as she wanted to turn around.

It was a bathtub, with claw feet like a beast's. The inside was rusted out and filled with brown rainwater, green algae floating on its surface. Something jagged snagged Frida's throat, and she swallowed it down. Here was evidence of other people. A person had abandoned this here.

What was she doing? She had to pee, and she only had two pairs of socks left. If Cal came to find her at the creek, maybe to talk, he would worry. And then, later, he'd be so angry. She had to turn around before she came upon other objects. Before someone stepped in her path with a weapon.

But first, she hiked up her dress and squatted next to the tub. She pulled down her leggings and peed. There was an atavistic relief to this, and her eyes watered from the pleasure. The end of the world couldn't take this tiny joy away from her. She was a dog, marking her territory.

Frida was here.

As she stood, pulling up her leggings, her dress falling back to her ankles, a sound caught her attention. Something like a crunch, like someone stepping on fallen leaves. She froze, that jagged thing rising in her throat once more. "Hello?" she whispered.

No answer.

Relax, she told herself, *it's nothing.* Couldn't be. But still, she thought she felt a presence not far from where she stood. Something, someone, was watching her, its breath shaping the molecules between them. She was breathing in that same air.

She stepped away from the bathtub. She would hurry back to the creek and then return to Cal. Nothing had happened; she was safe.

From behind a tree, another crunch. The sound came from her left, and she turned.

A coyote. It was standing there, watching her. It didn't look like the starving ones that used to skulk around L.A., desperate for a cat to eat, some garbage scraps. This one was well fed, big, with coarse brownish-gray fur that looked prickly to the touch. If Frida didn't know any better, she might think she'd run into a strange dog, tall and eerie eyed. It was so still.

A bird cawed above them.

Frida couldn't remember what she was supposed to do. Yell at it to go away, or step back quietly, or run like mad. This animal wasn't huge, but it could hurt her.

The coyote let out a rasping sound, its eyes arrowed into her, and Frida noticed an animal at its haunches. Something dead and small, a rabbit maybe.

"I don't want it," Frida whispered. She already had a hand on her stomach; already she needed to keep her child away from this. She could feel the fear growing on her like a skin, a mold. She could smell it.

The coyote pawed at its meat, rasped again. The dead thing had been torn down the middle and flattened like roadkill, limp and bloodied.

The coyote turned back to her, and Frida read its body, saw that it would pounce if she didn't get away. Above them the bird cawed once more.

Frida turned and ran.

On her way out of the forest, *away, away, away,* she grabbed every piece of clothing she saw, held them in her arms as if they could protect her.

———

"You look like a burglar," Cal said as she approached the house with the wet bag of laundry on her back.

"Help me with these," she said. She was still shaking.

"Sure thing." Cal grabbed the coiled rope hanging from the side of the house, stretched it to the tree across the yard, and hooked it. He kept looking back at her as he did so, as if trying to figure out what was different about her, as if she'd just returned from the beauty parlor.

"Are you all right?" he asked, when they were side by side at the line.

"No," she said.

"Please don't tell me you're still pissed at me."

Frida shook her head. "It's not that. Just now, after I finished the clothes, I went into the forest."

"You did *what?*"

"We need to find other people. They're nearby, I know it."

He didn't answer.

She squeezed his hand, hard. "Don't lie to me."

"Why are you being so careless?" he asked.

"Why are you being so duplicitous?"

He smirked. "Good word."

"I'm pregnant, Cal."

"So now you're sure about that."

"Would you rather I be sick?" She let go of his hand and pulled one of Sandy's old dresses from the laundry bag. It was wrinkled and cold.

He pulled it away from her. "Let me," he said.

"Just tell me what you know."

Cal closed his eyes, the dress in his hands. She could tell by the way his face scrunched up that she almost had him, that he'd do exactly what she asked. He wanted to tell her.

He almost dropped the dress, but she grabbed it. "I just washed this," she whispered. "Please don't get it dirty."

"I should have told you as soon as Bo told me," Cal said.

"What do you mean?" Frida thought of the coyote and the animal it had killed.

"Let's go inside," Cal said, and put a hand on her lower back.

6

They had packed two bags. Cal carried the larger camping backpack, and Frida had Bo's rucksack. *Rucksack:* that was Bo's word, and Frida said it made her think of a rugged country place, where men wore cowboy hats and called grown women *gals.* The rucksack was olive colored and dusty. After an hour, she said its straps bit into her shoulders, but she refused to let Cal carry it for her. She was trying to be tough.

If Bo was right, it would take two days to get to the fabled Spikes. They had agreed to get as far as they could the first day and get moving again early the second day, so as to arrive at the Spikes with the threat of darkness still hours away. They had the extra-large sleeping bag, and more than enough food, and even the flashlight, which Cal tried to convince himself they would only use if they had to. He doubted they'd be so disciplined. They'd both grown a little spoiled since moving into the Miller Estate.

Frida had been angry when he finally told her the truth. "How could you have withheld so much from me?" she'd yelled. But once they'd decided to go, she'd taken his hand and smiled, relief in her eyes. "We've got a plan," she'd said. That wasn't really true, though they did have a goal: they would present themselves to these strangers, these people who believed in containment. "And then what?" Cal wanted to know. Frida had no answer. Whatever happened, they would at least know who and what was beyond their land. She needed that, she said, and so did he.

For the trip, Cal was wearing Garrett's Official Pussy Inspector T-shirt. Frida had tossed it to him that morning when he was getting dressed. He'd laughed at first, and then, when he saw that she wasn't kidding, shook his head, which meant, he supposed, *I won't wear a dead boy's clothes.*

"If they ever spied on the Millers, they'll recognize it," she said. "At the very least they'll laugh. Anyway, you have to. It's your punishment."

She'd kept a straight face, but, later, when Cal shrugged the camping pack onto his back, Frida had barely managed to contain her laughter.

"That stupid T-shirt," she said. "Serves you right."

Before they left, he'd placed a hand on the front door and said a childish little prayer in his head, maybe to the house itself—*Please keep us safe. We'll be back soon.* Should he put his palms together, he thought, kneel by a bed? No need. The words alone comforted him, or just one of them: *Please.* He wanted to return to their little plot of land as soon as possible. The problem was Frida hoped to stay away for as long as they could manage. It was why, he realized, he'd taken so long to tell her the truth.

After his confession about the Spikes, they had spent two days discussing the inherent dangers of the trip. The difficult passage. "It won't be easy," Cal had said, feeling a sting of fear. "There might be rough bodies of water to cross and animals that come out at night. Who knows."

Frida had made up her mind. She told him it would be fine, that they just needed to get to the Spikes. Cal thought she was being naïve. "August must have had good reason to warn us to stay away," he'd said. She argued he was being a wimp. No one would get hurt; Cal's suggestion that she tie a pillow to her chest as a bulletproof vest was absurd, and she told him so.

"The worst that'll happen," she said, "is that they'll send us away."

Cal had said nothing as he tucked their pistol into the backpack.

That morning three days ago, Frida had left a rag hanging over the solar torch by the door, a signal that she'd gone to do some chores. Laundry, probably. But so early? He'd been a little relieved she was gone, actually. He felt exhausted by her, all her anger and questions. And yet, when he'd first turned over in bed and found her side of the mattress empty, it scared him. As far as he was concerned, Frida was the only person left in the world. He wasn't being poetic; it was a fact. And she might be carrying their child. She would become a mother. He couldn't lose her.

Cal had gone so far away from his own mom. That was the thought that rattled him until Frida returned. All he could think about was how distant he was from everything: from Ohio, from his dead parents, his boyhood. He wasn't even thirty, and already everything from his past was unreachable, not just Cleveland, but Plank, too, and L.A. The California he used to know. Sometimes Frida was so busy missing her own family she forgot Cal had lost one, too. At least her parents were still alive somewhere.

One of the last times he'd talked to his mother, she wanted him to come home for Christmas. "I'll show you the new short I've been working on," she said. Cal had been noncommittal, asked if she'd used actors, as planned, or something weird like Popsicle sticks. She was easy to distract if handed the right questions.

It was his second year at Plank, and he hadn't been back home since the day he'd left over a year before. He would have to fly, and the rising gas prices meant his father would have to sell his car to afford the ticket. More and more people were giving up driving altogether, and though Cal's father would still have the diesel truck, he'd have no backup. It was too risky. Cal was afraid, too, that his return flight would be canceled (that had become common lately), and he'd be stranded in Ohio. He could not stand the thought of missing his last semester at Plank.

When he talked to his mom again, it was really cold there. She said it was unrelenting, that ice was spiderwebbing across every window of the house, that there was so much snow she had a hard time opening the front door. She couldn't afford to pay the heating bill, even with the money his dad had lent her. Cal's parents had never been married, had never even been in love, and Cal's mother didn't like to lean on his father for a thing. Not that there was any more money to give her.

Every time Cal tried to call after that, the line was dead. How many times had he tried? Not enough. Plank didn't get the news right away. It wasn't until some other kid's parents learned about the storms, and thought to call the school in case any students had family in Ohio, that Cal knew for sure. The delay had almost been a blessing; he'd been spared the truth for as long as possible. Not that he hadn't worried. The week before the news arrived, he hadn't been able to get in touch with either of his parents, and he didn't know if they were okay, if they were alive, if Cleveland even existed anymore.

He should have asked to get off campus, to go online to find out, but he didn't.

When he learned that his mother was dead, he'd walked down the steps of the farmhouse, past all the boys with their books, past the cluster of Adirondack chairs that gave you splinters if you weren't careful, past the rusted tractor and the sleeping sheepdogs, and into the fields, deeper and deeper. He fell down in the mud, and he stayed there until he was too cold to move. Micah had come to retrieve him. "Come on, friend," he'd said, and pulled him off the ground. He wouldn't let Cal go until they were inside.

Cal's mother had probably frozen to death in the house he had grown up in. Even now, he imagined her wrapped in the old green-and-white afghan, shivering in her big sleigh bed, in her big bedroom that had once belonged to her parents.

He'd insisted on staying on at Plank. He went to class, he wrote his papers, and when the holidays came around he stayed on campus by himself. He celebrated New Year's Day in the stable with the horses, and he never talked about what had happened until Frida arrived for commencement. It'd been easy to tell her.

He'd forgotten how good it had felt, all those years ago, to spill his soul. She set him free, in a way, by listening.

When she'd finally walked up the path of the house carrying that big bag of laundry, he knew he'd confess everything Bo had told him. He wouldn't regret it, either, no matter where they were headed and what might befall them once they arrived.

They were officially in unfamiliar territory now; they had passed the bathtub filled with stinking rainwater a while back. "There it is," she'd called to Cal when its white porcelain side came into view, so smooth and stark against the trees.

Cal reached down to pick up a sock.

"What's this doing here?" he asked.

"I told you, I was doing laundry."

He waited; clearly, Frida hadn't told him the whole story.

"Just forget about it," she said.

"Where did you pee?"

She pointed her toe at the spot. "Voilà!"

Cal wanted to laugh, but he couldn't help but picture his wife out here, all by herself. An animal could've come upon her. She could have been hurt.

"If anyone tries to hurt you out there"—he swung his head in the direction they were headed—"I'll shoot them."

"I know you will."

"Are you feeling okay?"

"I'm great, why?"

"I mean, do you feel different?"

She paused, thinking. "I know everything will be fine, if that makes sense." She grabbed her breasts. "And my boobs, they're really sore."

"They are?" He put his hands on her chest. "They feel the same to me."

"I must be mistaken then," she said. "What do I know?"

"It was stupid to lie to you."

"We all have our secrets," she said.

They kept walking into the dense forest, where a few of the dogwoods were starting to change color. Frida allowed Cal to lead her, though he imagined she felt vaguely embarrassed to be following him blindly, as if he were her camp counselor. Bo had said only that the Spikes were due east, and already Cal and Frida's way had been obstructed by fallen trees and a wide river neither of them could have imagined and that they had to wade across, and the sound of animals was close enough to make Cal stop and reach for his pistol, one arm across Frida as though they were in the car and he'd stopped short at a red light.

They eventually rediscovered the tracks of August's carriage. Cal had been certain that would happen. From then on, they traveled more easily along his path. Cal thought August probably took a variety of routes; this one wasn't well trampled enough to have been used more than a couple of times.

"You don't trust August," Frida said from behind him.

"He doesn't trust *us*," he said. "Have you ever seen his eyes?"

"No. Have you?" He heard her fake gasp. "Are they made of glass or something? Or robot parts?"

Cal turned back for a moment. "Could be. I've never seen them either. That guy is always hiding something from us."

If they had still been new to the wilderness, the woods that surrounded them would strike him as identical to the ones they'd settled in. But Cal could see all the differences, however subtle: the space between trees, the light, the smells. It was incredible, to think this world had grown readable, as familiar to him as the street he'd grown up on. He couldn't fathom how strange it would feel to come upon these Spikes. Would he be too afraid to continue?

He began counting under his breath, *One-two-three-four,* again and again, a step for each number. He counted a little louder. These numbers would announce their presence.

Sometimes, as a safety precaution to scare away animals, he sang while they hiked; his father had loved Sinatra, and Cal could do a passable rendition of "I Get a Kick Out of You." Frida said she liked to imagine the bears swaying to his croon.

But this counting, it was different than singing. Something about the repetition, the way he could break the distance into these manageable parts, bolstered him.

He felt heat on his neck—a breath, a presence—and spun around. There was Frida, at his back, keeping close to him again.

"Hi, darling," she whispered, and like that, they kept walking.

7

Hours into their journey, Frida remembered something her mother had told her when she was a teenager. "I felt so confident when I was pregnant with you," Hilda had said. "And then it happened again, with Micah." She'd gone on to describe a peculiar peace that descended upon her with each pregnancy. As if, along with the necessary hormones and the double volume of blood swimming through her veins, a mother-to-be produced a reserve of courage for the life to come. Even naïveté could have a purpose. It was a survival skill, the same one that made a woman forget the pain of childbirth soon after it happened, so that she'd be willing to do it again someday. The species had to continue, didn't it?

Maybe Frida was feeling what Hilda had described. How else to explain how easily she pushed through these foreign woods, as if she would never be afraid again. She gave a secret nod to the coyote, hoped he'd eaten his kill and had taken a long nap after she'd run from him. Frida hadn't told Cal about the coyote, and she wasn't planning to. She deserved another secret from him. It evened the score.

At dusk they tucked themselves into what must have been a campsite for August and his carriage. It was a clearing just big enough to set up a tent and let the mare rest, drink water from one of the many nearby streams, maybe eat a bucket of oats. Was that what mules ate? Frida had wondered before where August had procured his animal, and if it slept in a stable somewhere, if it was offered a modicum of comfort and safety after each journey. Maybe Frida would finally find out.

Cal made a small fire while Frida unpacked their bedding and pulled out provisions for dinner. At the bottom of her backpack, rolled in a sweatshirt, nestled the turkey baster. She'd nearly forgotten about it. Her contraband.

She'd pulled it from the other artifacts after Cal had told her everything, and after she'd banished him from the house. She'd told him she needed to be alone to think, that he didn't deserve to share a home with her. Once she was alone, the plan was already sprouting in her mind: they would go find these people, and she'd offer the baster as a gift. This was how disparate civilizations were supposed to interact, wasn't it?

She hadn't told Cal about her idea. It was another secret she deserved.

She still had trouble believing that, for months, Cal had known about the insidious Spikes, had known that August traded with the people beyond them. Since hearing Bo's story, Cal must have conjectured about August. He might be from the Spikes himself, or he might be their leader. Cal must have reconsidered the Millers' death, too: Had these strangers wanted their friends dead? And why?

This is what hurt Frida the most: that her husband had bounced these ideas off the wall of his mind like the only child he was—alone, without anyone's input. He'd played with that tennis ball by himself, and he'd scuffed the same place on the wall again and again without any progress or relief. He'd acted as if Frida weren't there to help, or as if he wished she weren't.

They'd moved in together a few months after they began dating. It was a decrepit studio apartment in Hollywood, with a Murphy bed that came out of the wall. Their neighbors were either elderly or junkies, or both, always loitering out front or arguing with one another in the parking lot, and Frida and Cal would hole up in their place, lock the deadbolt, and tell each other about their lives.

Frida had told him how it felt to see her mother cry when Micah left for Plank. How she knew she'd never be enough for her parents. How neither of them expected much of her, how they believed her baking was silly—a stoner's hobby—and how, secretly, she agreed with them.

And Cal had told her how he couldn't stand to go back to Cleveland, even after they were allowing families of the deceased into the broken city, even though there was land that belonged to him. He didn't have the guts, he said. There were the Plank Chronicles, too. She could have recited the names of the animals there, the chores he did, the classes he took. He told her of his desire to carry the school's idealism into a world that maybe didn't deserve it.

Even though she knew it was arrogant to think this made them different from any other couple falling in love, Frida had believed that what they'd shared was more than what other couples gave each other.

But, now, she realized how silly she had been. She understood that these confessions, these stories about the past, were a rite of passage for any couple, clichéd but crucial, necessary to their survival. If she'd been with other men before Cal—not random one-night stands, or ongoing trysts with deli busboys, but real relationships—she might have known this.

She would have understood, too, that all the talking in the world couldn't give everything away, that a person was always capable of keeping secrets. It might have saved her from feeling betrayed by her husband here at the end of the world.

As twilight turned to night they ate beets and the remainder of their jerky in silence, the fire glowing orange between them, popping and hissing in that way that still delighted Frida, even after these two-plus years. She was relieved that she and Cal had been smart enough to travel during a gibbous moon so that it wouldn't be inky dark once the flames were extinguished.

Frida remembered how undark it had always been in L.A., the sky the green-gray color of something miasmic until well after midnight. She wanted badly to know what that sky was like now, if there was enough electricity to ensure that the city would remain bright and wasteful. Sometimes she pictured Hilda and Dada venturing out into the night together; in her mind they held hands.

After dinner, Cal tied the remainder of their food to a tree branch and then wiggled into the sleeping bag. He didn't ask her to join him; he had stopped requesting things of her since he'd suggested the stupid bulletproof vest. He probably wanted her to feel she was acting of her own volition, making her own choices, sharing in the difficult decisions of life. How thoughtful.

Frida didn't even pretend to have other plans: she got into the sleeping bag with him. He was her only shelter, and she wanted to be near him. The sleeping bag reminded her of their days in the shed; its slippery fabric smelled like mildew and dirt. If she let herself relax against him, she could enjoy this, the outdoors, the open space. The moon above them was the white button of a sweater, tucked halfway closed.

"I can grab the flashlight," Cal said. "If you want it."

She shook her head. "I'm okay."

"What do you think will happen tomorrow?"

"I have no idea." She didn't tell him that one moment she imagined pilgrim settlements and the next a high-tech world hidden in the brush: computer labs and electric toothbrushes, drivers texting from their hovercrafts. It was all so ridiculous, but in their Murphy bed in Hollywood she would have described each possibility to him in detail. She would have told him her biggest fear: that Bo had been fucking with him, that miles away there was nothing but more miles.

"They might kill us," Cal said.

"If you really think that, why agree to the trip?"

"Because you'd hate me otherwise."

His voice had turned hoarse, and Frida understood he was laying himself bare, making up for lost time, for past lies.

"I just want you to be prepared," he said.

"What? Prepared to die?"

He grabbed her leg under the covers. "No. But you need to remember that not everyone loves you immediately."

"Don't patronize me."

"They don't *want* us there, Frida."

"I need to be told that to my face."

"I know," he said. And then, after a moment: "Remember when we would go walking in your parents' neighborhood?"

Frida nodded. Cal knew she'd been ruminating on their courtship, on their young love. Either he could read her mind, or she was hopelessly predictable. Or both.

"Of course I remember," she said. They would go there to walk, because their own neighborhood was unsafe and ugly.

"I miss that," he said.

Frida nodded. They had only just moved into their apartment, and she'd missed her parents' house, her parents. It was her first time away from them, as pathetic as that was. It had been Cal who initially suggested they head there for a stroll.

They'd ride their bikes over so as not to waste gas and walk along the old familiar streets. "It'll make you feel better," Cal had told her, as if he didn't mind how dramatic she was being; they lived only twenty minutes away from the neighborhood she'd grown up in, and she was acting like they'd moved to the moon.

"Hard to believe those walks happened," Frida said to Cal now. "And here we are."

"It doesn't seem all that different."

Frida didn't answer because this was the root of the problem. Cal didn't feel any different—about her, about life—as he had all those years ago. For him, L.A. was the same as here. He'd been away from home since he was eighteen, and so everything was foreign, everything took some getting used to. She understood. Almost.

Frida was about to say good night, even though she was far from tired, when she felt Cal move closer to her.

They kissed, and he pushed himself against her, undoing the button of her jeans. She could feel by how desperately his tongue sought her own that he was afraid. Not of the night nor of the wildlife that probably surrounded them, eyes glowing yellow in the darkness beyond. It was tomorrow that frightened him. If these strange people welcomed them into their world, their lives would change. Again. Cal was trying to hold on to something. He was trying to hold on to her.

———

The next day was unseasonably warm. Frida could tell it would be hot before the sun had finished rising, and so she hadn't used any of their drinking water to wash up, even though her crotch smelled like manure. How sexy. She would be prudent with their drinking water, she told herself, she would squelch her vanity and her squeamishness. They would need water for the final leg of their trek, when they were tired and their mouths were sticky-dry, and Frida would be prepared. She could have bathed in a nearby stream if they weren't so eager to get going. Cal said there'd be more streams and bathing sources along the way, they were in a floodplain, after all, but she knew neither of them would want to stop.

Five hours into the trek, the dense forest cloaked the sounds of trickling water. The path was difficult and uneven, but Cal kept telling her they would keep east, no matter what. Hopefully they wouldn't have to cross any dangerous rivers, though here it was dank and still, and Frida wanted badly to dip her toes and dunk her head in water cold enough to turn her numb.

Finally, the woods gave way to a large field, carpeted with goldenrod and aglow with sunlight. Frida grinned. What a relief to be out of the forest! She could actually breathe.

"At last," she said, but Cal was quiet.

"What is it?" she asked, and he pointed into the distance.

She followed the line of his finger.

She'd been so enamored with the change in the landscape that she hadn't noticed the medium-sized school bus parked at the other end of the field, its yellow face yellower than the goldenrod.

"Look at that," Cal said, a smile breaking across his face.

Yellow. That unmistakable color. You'd think a world that was running out of oil, a land extinct of mountain lions and swordfish, would have also depleted resources of yellow dye. But no.

As Cal and Frida got closer, they saw that this bus wasn't like the abandoned bathtub—aside from a cracked taillight, it was in good condition, and it must have been used as a vehicle, as a transporter of people, not too long ago. It bore a fine layer of dirt and grit, and there was no license plate or any other identifying information on it. On the back were printed the familiar instructions: stop when red lights flash.

"Whose is this?" Frida asked.

"I wish I knew," Cal said.

She wanted to get inside it, but the accordion door was locked, and Cal didn't want to break in. "We don't want to give anyone a reason to hurt us," he said.

It was a good point. Besides, when Cal had given her a boost to peer into its windows, she saw that there was nothing inside: just rows of green-vinyl seats, waiting to be useful.

"It's really getting warm now," Cal said. He put a hand to the bus's side and removed it quickly. "We'd better keep moving."

By the time they saw the first Spike, rising sharp beyond the meadow they were crossing, Frida's shirt was heavy with sweat, and there wasn't much water in either canteen. They needed to save whatever little was left, Cal said, in case they were turned away.

From afar, the Spike was less ominous than Frida had imagined. At this distance, its surface glistened smooth in the sun, inoffensive as a sculpture in front of a bank. She had assumed, when she saw the first one, that she might feel an abrupt shock at its presence. She had pictured herself gasping or perhaps holding a hand over her eyes, her mouth falling open. But there it was, its existence undeniable, and it was immediately old news. She had longed to feel something stronger.

But then they kept going, up a small hill. Their journey had been on such flat land that it felt strange to be ascending. Exhilarating, even.

"Holy shit," Cal said.

Here, at this slight altitude, they had a better view of what lay beyond. It looked like the first Spike was leading the others. Frida counted twenty, and there were many more. They reminded Frida of the wind turbines east of the Bay Area: non-human structures that seemed, nevertheless, to possess consciousness and judgment. As if they were watching you.

"They're real," Frida said.

"Did you doubt it?"

"Didn't you?"

"Bo never could tell a good story."

That was true. A man with no past has little to narrate.

"Let's go see them," she said, and she broke into a run.

She felt as she had yesterday: brave and invincible, her body a machine whose only purpose was to follow her commands. The ground was uneven beneath her feet, and she skipped nimbly around rocks, asters, more goldenrod, a brown snake. She was getting closer to the Spikes, to whoever had built them. She kept running.

Her lungs were all right at first, and then they began to struggle. *Stop, stop,* they begged. She slowed. The Spikes were still far away.

She was bent over, gasping for breath, by the time Cal caught up with her.

"Please," he said, and put a hand on her back. "You'll hurt yourself."

When she was breathing normally, he took her hand, and they walked forward. Cal had once told her that the act that takes longer to achieve is often the more valuable one. The dinner cooked from scratch, the dress stitched by hand. The march into the unknown.

The closer they got, the more the Spikes resembled what Bo had described: they were at least fifteen feet tall, sometimes twenty or thirty, and each one was unique. They seemed to be made of metal, or at least pieces of them were, because they reflected the sunlight. They glinted in some places, maybe even sparkled. Some curved over like dying flowers, while others shot straight out of the ground—Frida had no idea how they were supported. They weren't smooth, as she had first thought, but bulky, uneven, and rough. She thought of the word corrugated.

The first one they saw up close was wrapped in chain-link fence and barbed wire, and it held all kinds of junk: a car bumper; a child's easel; an old plastic bottle, sinking into itself like a rotted bell pepper; and a walker for the elderly, tennis balls still stuck to its feet, gray instead of Day-Glo. The Spikes weren't spaced evenly apart as Frida had first assumed, and she thought she could see them inching closer and closer together the farther they walked. Did these beasts form a wall, a maze? How did anyone get out, let alone in?

"Are you sure you don't want to turn around?" Cal whispered.

"Of course I'm sure."

She knew what August had said, that the people who had built these things weren't afraid to use violence, but she nonetheless had the urge to keep going. The Spikes were ominous, casting shadows, their tips sharp, their edges serrated, but they were also beautiful. They changed the landscape, rendered it unfamiliar, even as they served their first purpose of protection from outsiders. She and Cal had ventured onto an unfamiliar planet, into an unidentified galaxy. Or they themselves had shrunk; they were ants walking among blades of grass.

Cal was holding her close, as if she might slide her bare hands across their barbed surfaces and perforate herself.

"I'm not stupid," she said. "Or fragile."

He smiled. "But you are pregnant."

If she was honest with herself, she could admit that Cal's gesture of protection turned her on a little. She had a passing vision of them getting naked right here, in the shadows of these terrible, stunning things. The Spikes were so breathtaking, somebody should.

Beneath the rusted wire of the next one, Frida saw junk she hadn't thought about in years: a lawn mower, a car battery, a stapler, a New Hampshire license plate, live free or die crimped and rusted along the last word. *Die.*

She wondered most at the stapler. Why that, of all objects? It was so small and ineffectual, but it could have been made into something else. A doorstop even, or a blunt object to throw at enemies. Its placement suggested wealth or profligacy, and, she had to admit, that was turning her on, too. In the last few years she had learned to sew up holes in her socks and underwear—*to darn,* for God's sake—and she was itching to waste something. Maybe these people understood that need and celebrated it.

She squeezed Cal's hand, and he squeezed back. She could tell he felt the same excitement, because he pulled her forward. "Come on," he said, his voice louder.

The Spikes reminded Frida a little of the Watts Towers: the sculptured junk, the imperfections. She had been only once, but Micah had begun going there regularly after he'd graduated from Plank. After some prodding on her part, he'd admitted that he went with friends from the Group. This was early on, before the Group had even hijacked the fund-raiser.

They stepped around another Spike. This one looked like it had been covered in papier-mâché before being wrapped with wire. It resembled an unfinished and nefarious piñata—instead of candies it would spill empty, yeasty wine bottles, splintering table legs, an old espresso machine. Once, in L.A., Frida had seen a barista apply red lipstick using the reflective surface of her coffee machine as a mirror. It had made Frida's day, the way objects could be remade, given a new and unexpected purpose.

"Micah would have loved these," she said.

Cal raised an eyebrow, and she knew she couldn't say what else she was thinking: that the Spikes were magnificent, proof that the people who had built them were magnificent, creative and daring, and threatening, too. But it was only a taste of threat, a dash of it, for flavor. She knew if she told all this to Cal, he'd say she was being naïve again, that she had too much faith in people and in their capacity for joy and art.

They kept walking. There was no doubt they'd gone farther than Bo and Sandy had dared, for the walk became more maze-like and challenging, nothing like what Bo had described to Cal. A few Spikes had been built so close together, they were impassable, and Frida and Cal had to navigate around them, doubling back until they found a wider path. Frida was thankful each Spike was so specific, each one its own landmark; otherwise, they would have no way of knowing if they were moving forward, backward, or in circles.

"This is like a video game," Cal said at one point.

"Like your mom ever let you play a video game," Frida said.

When they hit another wall, six Spikes so close their necks intertwined like swans', Frida felt her first pang of fear. She tried to ignore it, but she couldn't. They might be stuck in here, she thought, panic winching her throat closed. How long did they have to walk? Would they ever get there? She shook the questions away, tried to play it cool as Cal pulled her left, his forehead wrinkled in concentration. She wanted to make him laugh, wanted to maintain a clear mind. There was an end to this maze. The trick was not to freak out. She kept looking to the ground, at the dead grass beneath her. The land was flat here. They could have been on an old soccer field. In the last hour, the entire world had begun to feel man-made, not just the Spikes. This should have made her hyper with expectation, but now she felt as Cal did: hesitant, suspicious.

She had just asked Cal for a sip of water from his canteen—screw the rationing; why hadn't they followed that sound of water, a ways back?—when she heard a sound like knuckles cracking, like the twist of an old man's spine. She stopped.

"What?" Cal whispered.

"Listen," she said, and thought immediately of Micah, of his last word, which she had exiled from her vocabulary after his death. Cal must have noticed the word, too, because he had turned pale.

The sound again. Frida imagined someone, a man, hiding behind one of these many Spikes. He was cracking the knuckles on his other hand, one by one.

"Who's there?" Cal called out. His voice was steady, and it gave Frida confidence.

"We've come for help," she yelled, trying to sound as calm as he had.

Nothing. They waited. And waited.

A tightrope of anxiety strung itself sharp across Frida's body. Her poor baby. This feeling, if it remained, would ruin him. He'd come out of her trembling.

Frida looked up. The Spike next to her was the tallest one yet, and at its top sat an orange traffic cone, bent over in defeat as if a big rig had run over it.

She was still looking at it, wondering if it was a signpost for these people, when she heard a whistle like a catcall, and a man stepped from behind a Spike a few feet away.

He was more of a boy. He couldn't have been older than twenty, and he was thin, small framed. Frida was shocked by how normal he looked; he didn't wear a gown of feathers or chain mail or a silver space suit. He had on a dingy white T-shirt; the elastic around the neck was shot, so that it ruffled flaccid at his collarbone. Brown corduroys, the hems frayed. His sandals looked like they'd been made from tires.

"Go away," he said. He carried a rifle, but he held it against his thigh, as if he had no intention or desire to use it.

Cal stepped forward and put out his hand slowly. "Please. My wife and I have come to find out who's out here ... We've settled so close to you."

The man looked Cal up and down. He did not take his hand.

His hair was dark brown and pulled into a scraggly ponytail. His eyes, also brown, were clear and focused. There was nothing shifty or unpredictable about him, Frida realized.

"Go away," he repeated.

"No," Frida said. "Not until we get some help."

He shook his head. "You need to leave. I can give you water, but that's all."

Frida put a hand to her chest. "That's it?"

The man raised an eyebrow and turned to Cal. "What's with the shirt?" he finally asked.

"Pussy is a kind of mushroom," Frida said.

Cal blushed, and the man just looked confused.

Frida looked away from his face and to his hand, the one that was wrapped around the gun. It was calloused and dry, his fingernails caked with dirt. The man lifted his left hand to scratch his smooth cheek, and Frida saw that his fingertips were peeling, almost chapped.

Tattoos had been so common by the time they'd left L.A., Frida hadn't noticed the ones on this stranger at first, but then they were all she could see. A single blue party balloon floated across the inside of his wrist, and what looked like an octopus tentacle, suction cups and all, peeked from his shirtsleeve. There was an old-fashioned anchor, too; it was no doubt ironic, this kid hadn't been on a ship in his life.

The line of anxiety that had been strung so taut across her snapped; she thought she could feel her baby, falling from that uncomfortable balance, back into an easy sleep.

"We don't want water," Frida said.

"Look," Cal said. He laughed. "I was almost about to say, *Take me to your leader*, but that makes me sound like some kind of alien invader."

"Please," Frida said. "We just want to meet you all and get some help."

"Not going to happen," the man said, shaking his head. Frida saw that his ponytail had been secured by one of those hair ties with two red plastic balls, as if he were a little girl.

"We believe in containment," he said.

"So we've heard," Frida said.

The man raised an eyebrow. "Is that right?"

"We know August," Cal said.

Recognition passed over the young man's face, but he said nothing. He was holding the rifle tighter now, as if more conscious that the weapon was at his disposal.

Cal wasn't giving up. "I'm Cal." He put a hand on Frida's back. "This is my wife."

"My name is Frida."

It was if she'd said, *Open sesame*. The young man looked up suddenly. "Frida?" He shook his head as if he were emerging from a cold pool of water.

"That's me," she said.

"Frida? That's your name? Really?"

"Why does it matter?" Cal said. "You don't know us."

The young man bit his lip. "You better come with me."

8

The kid's name was Sailor. "That's my real name," he explained, after Frida asked if he'd been christened that by friends. She had gestured to his forearm, and Cal saw that it was tattooed with a solid black anchor fit for Popeye. Sailor shook his head, told her his parents had been whimsical people. "Child as art project, that kind of thing."

Cal wouldn't have noticed the tattoo if Frida hadn't pointed it out. But she had probably recorded everything about Sailor; she probably liked his narrow shoulders and his nervous bravado and the way he just kept saying, "Follow me," whenever Cal asked where they were headed. Frida was obviously smitten—she couldn't hide it.

Or maybe she just felt protective of the kid, her maternal instinct kicking in. He looked so young. Sailor had told them he was twenty-two, but that seemed impossible. He reminded Cal of certain first-years at Plank who ate and ate and never gained an ounce, who had yet to grow chest hair or even a passable goatee. In other words, a Planker like Cal had been. He hadn't been malnourished when he arrived for college, just young, boyish.

Cal was relieved to have a guide, at least. Someone who understood these Spikes and the labyrinth they formed, who wasn't intimidated or enamored by them. The latter was Frida's problem; she walked around each one with awe, as if the Spikes were brilliantly rather than sloppily constructed, as if they were any better than the découpage and found art projects his mother had done with her friends every other Tuesday night when he was a teenager.

Her *salon,* she'd called those get-togethers.

Cal was impressed with how mazelike the Spikes were, though: how they could confuse and terrorize a stranger, keep him out, force him to give up, go home. He longed to see the intricate route from above. He wondered if together the Spikes formed a beautiful design, like a crop circle. Or maybe a word. *Boo!* Or a phrase. *Crown of thorns.*

The words had shot across his mind as if from Sailor's rifle, catching him by surprise. *The crown of thorns that surrounds the city of God.*

"What's that?" Frida said.

He hadn't meant to say the phrase aloud. If he remembered correctly, the quote was about Rome, about the shantytowns that encircled the city. He couldn't recall who had said it.

"Pasolini," Sailor said. He was walking just ahead of them and turned to smile.

How had this kid known that reference? Because he didn't want to betray just how impressed he was, Cal simply nodded at Sailor and kept walking.

"What are you guys talking about?" Frida asked.

"Famous words by famous men," Sailor said. "That's all."

He led them around another series of Spikes, and then another. For a moment it seemed they were doubling back unnecessarily, and then Cal realized they had done so to avoid a veritable castle wall of Spikes, built so close together their trunks kissed.

And then Sailor pointed to the ground and said, "Careful."

Before them, in various places, pieces of glass protruded from the dirt and grass. Cal had read about places in Latin America where they lined walls with bottle shards to keep people from climbing over. These pieces of glass, which he and Frida tiptoed warily around, had clearly been placed for the same purpose: to slice people's feet, to maim them or, at least, wreck their last pair of shoes.

Because this trail of glass required that they keep their eyes to the ground, Cal stopped walking and looked up. He wouldn't follow the maze's implicit rules just because he was afraid of a torn-up heel. These people, they didn't want him to take stock of location, perspective. Well, he would.

The sky was blue but hazy in the way it got when it was hot, and only one crimped ribbon of cloud interrupted the solid color.

From behind, Frida tapped his shoulder. "Go on," she whispered. She thought he'd stopped because he was afraid. He kept moving.

Sailor seemed almost giddy as they reached the edge of the maze, and Cal wondered if anyone had ever done what he was about to do: bring outsiders inside. Sailor wore a stupid saggy grin on his face, like he'd just won a first-grade spelling bee. Cal could hear Frida behind him, her breath loud and shaky. She only breathed like that when she was nervous.

Cal trusted Sailor; the kid was obviously guileless, and he didn't seem cunning enough to do them any harm that Cal wouldn't see in time to prevent. But maybe Sailor was unknowingly leading them into danger. His compatriots might not agree with his choice to accept two more people. The only person who could have been talking about them was August, and Cal couldn't imagine what August had said that made Sailor suddenly so welcoming. One second he'd been telling them to get lost, and the next they were the guests of honor. And it was Frida's name that had been the magic password. Cal didn't like that. He didn't like that he didn't like that. It meant, if he had to be honest with himself, that he didn't trust his wife.

All at once, sudden as a hiccup, they reached the end of the labyrinth. Sailor had led them around one more Spike, and it was over. They were back on flat ground at the edge of a field. There were no trees in the distance—they must have been razed. Were it not for the Spikes behind them, it might feel like they were standing at the center of any large suburban parking lot, and maybe they were. Cal looked up once again at the wide-open sky and remembered his one and only trip to Cedar Point with his father, before the amusement park had been shut down. At the end of the day, his father had forgotten where they were parked, and they'd ridden the lot tram for forty-five minutes until they finally discovered the truck, tucked into a line of minivans. Now, Cal rubbed a foot along the grass before him. It was dry and striped with brown, the kind that sprouted like tinsel out of asphalt. If left to prosper, it could grow into a field. Cal shuffled at the grass, and sure enough, it gave way to a patch of concrete beneath.

"Don't move," Sailor said.

Across the field loomed a wooden platform. It was a lookout tower, the kind prison guards watched from, but rudimentary, as if it had been built in a rush. But it was certainly in use: at the top stood two figures, binoculars in front of their faces.

Sailor stepped forward and waved his arms, the one with the rifle in it a little lower and slower moving than the free arm. Frida laughed. She had probably noticed that the gun was too heavy for the poor kid. Cal winked at her.

Sailor began moving his arms in a choreographed sequence that must have meant something to his compatriots. He looked like a majorette, and Cal felt transported home, to the Midwest, with its flag girls, its chilly autumn evenings, and what felt like the whole world preparing for winter. But it wasn't the whole world, because Frida didn't have any of those memories; she'd once told him that they didn't make those kinds of nights in California. Or those kinds of girls.

Cal watched as the men in the tower above them waved their arms back at Sailor. One of them reached behind his back, and Cal stepped in front of Frida. But the man had pulled out not a weapon, but a whistle, and he blew it three times, each note long and piercing.

"Does that mean 'intruder' or something?" Frida asked.

Sailor laughed. "Nah," he said. "It means I've come back, with two strangers. He's telling the others. But don't worry, it's not the panic whistle."

Frida nodded. "I like you better when you're forthcoming." Was she *flirting?*

"Just wait till he sees you," Sailor replied.

The whistle blew again, five quick bursts, and Sailor nodded, flung his arms up once more. "Follow me," he said. He had turned official once again.

As they crossed the field—*A parking lot,* Cal told himself—Sailor moved his rifle so that he was holding it diagonally across his chest. This was probably how he was supposed to carry it; he'd get in trouble for letting down his guard, even if it was for Frida. Maybe she was their god.

Two men jumped from the bottom of the tower's ladder. One was about Sailor's age, Cal guessed, but broader in the shoulders and bearded. The other looked about forty. They wore ripped-up jeans and old T-shirts. The older guy's had a picture of the Olympic rings.

All at once they were running at them, like soldiers.

"Hey now," Cal said, and once more stepped in front of Frida.

"Be cool," Sailor said. Cal didn't know if the comment was directed at him or the others.

The Olympian put out a hand, covered in cuts, thumbnail black and warped. "Your bag."

Cal turned and saw that Frida had already handed the rucksack to the other one. "I need it back," he said.

"He's just searching it," Sailor said. "Relax."

The Olympian actually said thank you when Cal handed over the backpack, which Cal appreciated. They could all be civil.

"You should also know," Cal said, "I'm carrying a pistol." He put up his arms immediately, so that it wasn't perceived as a threat.

The Olympian nodded, and Cal pulled the gun from the back waist of his pants. Sailor took it with a smile. "I knew we could trust you."

"What's your name?" the Olympian asked. "I'm Peter."

"Cal." He grabbed for Frida's hand. "This is Frida."

"Frida!" Sailor repeated fiercely. Peter socked him in the shoulder.

The other man was named Dave. Cal was glad they didn't all have cutesy names like Sailor. Dave had chosen to kneel and go through Frida's rucksack item by item. Cal turned just as Dave was pulling Frida's shirt out of the bag like a magician's endless scarf. He shook it out and, satisfied that it didn't hide any knives or bombs, tossed it to the ground.

"Come on," Frida said. "Really?"

"Let's not get too security guard on her, Dave," Peter said.

Dave looked up at Peter and scowled. Unlike Sailor and Peter, he had shorn off his hair, and his scalp was pink beneath the blond bristle. What an idiot, Cal thought. Not even a hat.

Dave was rooting around deep in the rucksack, his brow furrowed. Cal imagined him as a former mechanic, diagnosing car parts.

"What is this?" Dave said.

Frida stepped forward. "Please, it's meant as a gift."

From where Cal stood, it looked like Dave had pulled out a discarded page from a magazine, wrinkled as a pirate's map. It was just paper, but it was wrapped around something.

Frida looked at Cal, and then to Sailor. "Please don't."

Dave stood up as he unwrapped the paper. Cal saw it was old wrapping paper, shredded at its edges, from a Christmas long ago. Was it possible that he recognized those anthropomorphic gingerbread men with their demonic red eyes and meaty fingerless hands?

Cal turned to Frida. "What *is* that?"

"You don't even know?" Sailor asked.

Frida tried to take the package out of Dave's hand, but he stepped away from her.

"I told you," she said. "It's a gift. Please just leave it be."

"If it's a gift," Dave said, "then I'm excited to open it."

"Maybe it's for you," Sailor said to Cal, smiling.

Dave unfolded the paper and pulled out a turkey baster. "This?"

Frida nodded. "I don't know who it's for. It's an offering. I guess."

"Where'd that come from?" Cal said.

"One of my artifacts," she said, under her breath.

"What are those things called again?" Sailor asked.

"It's a turkey baster," Cal said.

In another life, this would have been any other piece of kitchen equipment, though rarely used; his mother had only trotted theirs out on major holidays, and once, as part of his fine arts credit, she'd had him draw it, first in pencil, and then with charcoal. He didn't remember owning one himself. And this one was new and fancy, its cylinder made of glass. It still had its tag.

In another life, Frida would not be in love with this object, but he could tell, by the heat that colored her neck pink, that it meant a lot to her. More than it should.

He imagined taking the baster from Dave and cracking the cylinder in two over his thigh.

"We've got one of these already in the kitchen," Sailor said, taking the baster. He held it up to the sky. "But that one's cloudy and made of plastic. This one's a beaut!"

Peter took the baster from Sailor and handed it back to Dave. "Wrap it back up. Let her choose who to give it to."

Dave folded the paper around the baster and shoved it along with her sweatshirt back into the rucksack. "We'll keep these bags for the time being," he said, and Cal was about to protest when he saw that Frida was merely nodding. She looked at the three men, then at the ground, once at the sky, and back at the Spikes. She was looking everywhere but at Cal.

All at once, he pitied her. His dear wife. When had things gotten so bad? He remembered something his father had once told him. "People get sad." It was true. Maybe sadness was where they were all headed.

"Let's go," Peter said, and blew his whistle three more times, the same message as the first. After a pause, he blew it once more, quick and sharp. His whistle was a train station announcement, a grandfather clock, an emergency broadcast system.

Cal and Frida were led onward. As they passed the lookout tower, Sailor grabbed one of its wooden girders and spun around it as if it were a telephone pole and he were the star in a musical.

"Calm down," Peter said.

"How can I?"

Dave grunted.

Past a few spindly trees, they reached what looked like a kiosk at a movie theater or a small visitors' center. Cal was all at once back at Cedar Point, the line to buy tickets not as long as it should have been. He remembered the way his father had leaned into the windowed booth to give the uniformed woman cash for two tickets. On their way out, a different woman had stamped their hands with fluorescent ink that smelled of lemon cleaning spray. "So's you can get back in," she'd explained.

The windows of this kiosk were covered with foil, and he couldn't see in. By the blisters in the paint and the vines crawling along the sides, it was clear that its original purpose had long been discarded.

"What used to be here?" Cal asked.

"On the Land?" Sailor said.

Frida smiled. "The Land? Is that what you call this place? I guess that's a little more inviting than the Spikes."

Even Peter laughed at that one. "Keep walking," he said. "It'll become evident."

The Land. Appropriately vague, vaguely philosophical, a tinge poetic. Cultish, maybe. But, at least, if the place was named, it meant its citizens wanted to stay here, make it into something worthwhile.

The kiosk led to one long dusty road. It was a main street, the kind Cal saw in his mind's eye when he imagined fledgling mining towns, new settlements of the Wild West. The world of children's history books, adventure stories. This had once been the kind of place where men settled arguments with bullets, and the rare woman flashed a leg at the saloon; a town that had probably been forsaken in a day, and easily: all of its inhabitants leaving at once, driven away when the wells ran dry or when the gold ran out. Or after an earthquake had unbuckled its foundation, scaring the easterners silly.

"A ghost town?" Frida asked.

That same dopey grin spread across Sailor's face. "The ghost of a ghost town."

Cal raised an eyebrow.

"It was built in the 1800s and then reopened a few decades ago as a place to wear a cowboy hat and imagine the past. Closed down about twelve years ago, after it cost too much to keep it running. I guess people stopped driving out of their way to leer at a bunch of abandoned buildings—too costly, and the roads just got worse and worse. We still have some pamphlets, though, from when it was open as an attraction. I can show you later."

"Sailor," Peter said, and the kid stopped talking.

The road was lined with sagging wooden houses. There were two larger structures farther down. A few had plank walkways leading to their doorways—some of the doors had long blown off. One of these openings was covered by a large animal hide; it looked coarse and crusty, and it flapped in the breeze.

Near the kiosk stood a single brick wall, freestanding and crumbling, uneven. The other parts of the building must have collapsed and then dissolved. Or, more likely, been reused. The houses on either side of them were part ghostly and antiquated and part rehabbed. The two parts didn't match. A few of these extensions looked like the house he and Frida lived in.

"The Miller Estate," Cal whispered to Frida and gestured to the house to her left.

Peter looked like he was about to say something, explain, but Frida grabbed Cal's hand and said, "Oh my God." She was looking ahead, down the road.

People. There were people.

They were emerging from the houses and bigger buildings, one of which was a large church, with its steeple wrapped in barbed wire. One more Spike.

Cal pulled Frida to him and kissed her cheek, swept her knotted hair out of her eyes. She smelled of sweat, like the muskiness of the tea she used to like to drink in winter and the canvas of Bo's rucksack. They would soon be carried away by this tide of strangers, and he didn't want to lose her. Or was it her familiarity that he didn't want to lose?

She smiled at him, but her gaze went right back to the people coming toward them. Cal couldn't blame her; it was mesmerizing to see such a population. There must've been forty or so. All those faces and bodies. From afar, they looked like Peter, Dave, and Sailor. That is: human. Like Sailor, they had on normal street clothes, if a little torn up: T-shirts, jeans. Work boots or sandals, a few maybe barefoot. There were women, too, and some of them wore long dresses, the kind Sandy had been partial to. Thank goodness there weren't only men, Cal thought. He didn't want to enter a world like Plank. He didn't want Frida to be the only woman.

As they got closer and the people's faces began to differentiate themselves, in all their unique ways, Cal felt light-headed. It was the same feeling he'd experienced his second year at Plank, when he'd gone to a street fair three towns over and seen so many people he felt drugged by the newness.

He'd gone with Micah and a first-year they hardly knew. And that guy's cousin—a real live girl. It was Micah's Toni, though at the time, Cal had called her Antonia in his mind because her nickname sounded too much like a guy's. She was three years older and lived in L.A.

No one was supposed to sleep on campus except for Plankers,

but visitors came so rarely, and female ones never, that everyone let it go. Besides, she'd come with a trunk full of groceries for the students: Cheez-Its, celery, and apples from New Zealand, almost too expensive to eat. Her grandmother had money, though she claimed she never talked to her, that she'd run away years before. After one night on campus, Toni said even the quiet itself was boring, and so she'd driven her cousin, Micah, and Cal to the strange fair she'd seen signs for on the drive over. Cal wasn't sure how he and Micah had smuggled themselves into that car; the other Plankers had been so envious.

The fair was in a suburb. An exurb, really. Wasn't that the word? It had been so clean and antiseptic, Cal thought now. It was transformed into a Community soon after.

"How stimulating," Micah had remarked as they passed beneath a banner painted with a garish rainbow, and Toni had laughed, thrown her head back as if she wanted him to slice open her neck. She'd been smitten with Micah immediately, the lucky bastard.

There were packs of people everywhere: watching juggling acts, eating corn on the cob, dancing to the music of a leather-vested fiddler. Cal wasn't used to so much diversion, so much information. The colors, the noise, the sugar, and the salt. The fair left him giggling like a stupid drunk.

That's what Cal wanted to do now. Giggle. He wasn't happy, and this wasn't funny; he was just overwhelmed. For months it had been just the two of them, Cal and Frida; even when the Millers were around, and even with August's monthly visits, they negotiated a very limited universe. The same trees to count and admire, the same gardening routines. Suddenly everything and everyone were new. No wonder Frida didn't want to look at him.

He would look at her, at least.

She was bobbing her head like a pigeon, taking in the sights, and smiling with her mouth closed. She was trying to appear kindly, he realized. She wanted these strangers to like her. As far as he could tell, they would. Most of the men seemed around their age, or younger, like Sailor, and none of them walked with the defensive stance he'd expect from a culture that did not allow outsiders.

The women were more hesitant, hanging back, and maybe a little older on average: in their forties or early fifties. One of them wore her hair in a thin braid down her back like a second spine.

Frida put her hands to her lower belly as she walked. She was thinking of their child. He could see she already felt safe, protected, that she was fly-casting them into a future in this world. She was being naïve. Again. They would have to talk, and soon.

Dave had left their little posse. Maybe he'd been pulled back to the lookout tower to finish his shift, but Peter and Sailor led the way through the crowd, which parted to let them pass, just as Cal had imagined it would. The people, up close, were so varied: heavy browed or not, ugly or cute, plain or strange and uneven looking, long or pert nosed, fair or olive skinned.

There was a woman with a stripe of gray in her curly black hair, thick red suspenders holding up baggy corduroys. She had a rag in her hand, and when she smiled, she was missing a front tooth. Cal reared back. He couldn't help it. No doubt this woman had lived out here for a long time.

Cal accidentally brushed past a heavyset man, about his own age, who stepped back with a sneer. That was the only rudeness he encountered, and Cal couldn't hold a grudge: he and Frida had invaded their space.

A guy with dreads so blond they were almost colorless

muttered to his friend, "Would you look at that." He pointed to Cal's chest.

Peter turned around and began walking backward. "You do a lot of mushroom hunting, Cal?" he asked.

"I guess."

"Your shirt," he said.

"You knew the Millers?"

Peter frowned. "Who?" He spun back around and kept walking.

People had begun to come up to him and Sailor. They were asking the same question— *Who? Who are they?*—and pointing at Frida and Cal.

A woman stepped into Sailor's path and asked, "Who let them in?"

"You'll see, Pilar," Sailor said. He cupped his hands around his mouth and yelled, "Someone get Mikey!" The woman in the suspenders took off running down the path.

"Who's Mikey?" Cal asked. "Do you mean August?"

Peter stopped walking and turned back to Cal. "You better hold her."

"Me?" Frida said. "Are you about to sacrifice me or something?"

Peter kept his face serious. "Relax, okay?"

At the end of the path loomed the other large building, not as wide as the church but taller if you didn't take the steeple into account. Like the other original buildings, it was built of wooden planks, but it looked somehow sturdier than its counterparts. Still, the windows on the upper floors were empty holes, most of them covered with cloth. A wide front porch stretched on either side of its entrance, and a slanted awning offered shade and a place to rest; turned-over crates acted as chairs, as did a weight-lifting bench. No one had added on to

134

this building, and from this distance it didn't look too decrepit. Fetched from another era, it had probably once been a general store, the town's unofficial beating heart. It most likely still was.

A man stepped out of the building. They weren't so close that they could make him out, but Cal could tell it wasn't August: this guy was white and burlier. Long brown hair stuck out from his large-brimmed straw hat. His beard reached his chest. If it weren't for his jeans and green Polo shirt, he could've passed for Amish or a hippie.

"Is that Mikey?" Cal whispered to Frida.

She stopped walking. Her arms hung tense by her sides, and Cal saw that she was clenching and unclenching her hands into fists. But the fists, they were weak, not fierce. She began to shake.

"Hey," Cal whispered.

She took fast steps toward the end of the path. She stopped again. She was craning her neck forward, as if to get her eyes closer to the sight before her. Cal looked back at the man, to try to see what she was seeing.

"No, no, no, no," she began. Her voice squeaked out of her in little high-pitched bursts.

"What is it, babe?" Cal asked.

Peter was by their side. "Relax now," he said to Frida. "It's okay."

Sailor was bouncing in place, his eyes wide and glistening.

"What's going on?" Cal asked. He felt his whole body go cold. But why?

An inhuman sound emerged from Frida, full of sorrow and giddiness. It was as if she had moved beyond words. She staggered forward, and Cal tried to follow her, but his body wouldn't move. He felt trapped; all he could do was watch. He didn't understand. What was wrong?

"Is it?" Her voice came out as breath. "Is, is . . ."

The man at the porch was standing steady, just waiting for Frida to meet him. He wouldn't meet her halfway. He opened his arms wide.

Cal's heart beat in his eardrums. *Mikey.*

No, not Mikey. And not Mike E.

Mic. E, as in Micah, as in Micah Ellis.

Micah.

The world slipped sideways for a moment, Cal's stomach lurching with it. He leaned over and vomited into the dirt.

"Easy now," Peter said, and patted him on the back.

Suddenly there was a canteen of cold water at Cal's mouth. But he could not swallow. He forced himself to stand. Frida was just a few feet away from the man who looked like Micah. She was weeping, hiccuping.

"You," she was saying.

The man who looked like Micah held his face steady, as if trying not to betray whatever lay beneath his placid expression. Cal saw it in his eyes; they were darting over Frida's face, taking in the ways she had changed and aged and the ways she had remained the same and would remain. The man stepped forward finally and took Frida into a bear hug. She collapsed into his arms, her legs giving out. He held her up.

"Yes!" Sailor cried, his arm pumping in victory.

Cal was shaking as Frida had been. He felt like vomiting again, but his stomach was empty. He held on to Peter's canteen and willed himself forward.

"Micah?" he tried.

The man looked over Frida's shoulder as he held her.

"Is that you, California?"

9

He smelled the same. She hadn't hugged him for years; even when he was alive, they barely touched, but now she couldn't let go. That smell: what was it? Pajamas worn until noon, and potato chips, and the leather band of their father's favorite watch, and the baby detergent their mother never stopped using, and his old room, the window never open, the blighted avocado tree blocking views and voyeurs alike. Her brother, his smell.

She couldn't stop embracing this ghost. A ghost in a ghost of a ghost town. Ha. It was a word problem, a riddle, a mirror inside a mirror inside—of course he loved that.

Micah. Her brother was alive.

His shirt was the color of a tennis ball, and she was imprinting its insignia of a man on horseback onto her cheek. She was pushing her face so hard into the ghost's shirt it was like she wanted to graft the fabric onto her own skin. Not a ghost. Her brother. Micah. He had on such a stupid shirt, and a theatrical beard and a farmer's hat, and he was breathing deeply, as slowly as a bridge rises to let ships pass. Was she the ship?

"Hi, Frida," he said into the top of her head, so quietly that no one else heard. And when he pulled away to greet Cal, she almost fell.

Nearly every time Frida tried to open her mouth, the words clogged in her throat, and she stood there dumb and struggling. They'd been following Micah as he led them back up the path. *Is this real?* she wanted to ask. *Are you really here? Will you let us stay?* Instead of speaking, she wandered the Land like a child in a picture-book world: blue sky, brown dirt road, yellow sun, her mouth a flat black line. No text.

Cal was trying hard not to roll his eyes at her. Not hard enough. She could tell he wanted her to get her shit together and help him figure out what was going on, but she was incapable. Her legs still felt rubbery when she walked, and she didn't think she'd be able to speak ever again. Her hands seemed to belong to someone else, and her mind kept returning to that first sight of Micah, of his green shirt and his long beard.

Her brother!

Cal had already asked him half a dozen questions. Micah hadn't answered many of them, but that didn't keep Cal from trying.

"How are you alive?" he'd asked as soon as he could, even with everyone watching.

"It wasn't a resurrection," Micah replied, "if that's what you're asking."

"I wasn't," Cal said, and Micah laughed.

Cal said nothing. He looked at Frida, as if pleading for her assistance, but she looked away. The truth was her brain was still playing catch-up. Micah was alive. Her brother wasn't dead. Micah was alive.

She felt a little sick.

When Micah stepped away from her to greet Cal, Frida had slumped forward and caught her balance just in time.

"So you two are still together," Micah said.

Cal raised an eyebrow. "Don't act so surprised."

"Oh, never," Micah said, and winked at Frida. "First things first: you guys need to wash."

"We do?" Cal asked.

Micah smiled. "You might've seen a little building on the way in—a kiosk-type thing. We call it the Bath. In there we've got antibacterial soap and talcum powder. Even though we also have outdoor showers, you'll probably feel more comfortable with a little privacy."

Frida had felt a rush of relief then. Clean. They would get clean.

"I'll meet up with you after that," Micah said.

He was already nodding at Sailor, turning away from her. Frida couldn't believe he was leaving them again, and so soon. He very well might disappear. He'd done it once before.

Her brother looked at her and smiled, gently. "You won't be gone long," he said, and because Frida couldn't speak, she followed Cal and Sailor.

Sailor led them to the Bath. Inside, there were two shallow plastic tubs, the size of foot baths, and two plastic chairs that looked vaguely medical; Frida imagined they'd been used in hospitals, for the sick or elderly, people who needed to sit down while showering. An array of products awaited them on the built-in counter where an employee must have once peered out the ticket window: the soap and talcum powder Micah had mentioned, various creams and lotions. Even a bag of disposable razors; Frida's heart quickened at the sight of them.

"Where did you get all this?" Cal asked Sailor.

"I'll get you the water," Sailor said, and ducked out.

He returned with a bulky canvas bag slung onto his shoulder and two big buckets of water, which he poured into the tubs.

"Sorry for the temperature," he said. "You're not on the bathing schedule yet—so I don't have anything sun heated for you."

"It's fine," Frida said. It was the first complete sentence she'd said in an hour, and she couldn't help but feel triumphant.

"Wash your feet, pits, and genitals," Sailor said. "Use the antibacterial soap in the back row, and any of the creams, if you've got rashes or something." He held out the canvas bag, and Frida leaned forward to see what was inside of it.

"They're clean clothes," Sailor said. "I think they'll fit."

"They won't be necessary," Cal said. "I can make do with the clothes I already have."

Frida crinkled her nose and put her hand out. "I don't know what's wrong with Cal, but I'm happy to have something clean to put on. Thanks, Sailor."

Sailor smiled and handed her the bag. "Anyway, I'm glad Mikey said something right off. You guys are rank."

"Sorry," Cal said. "August never has this kind of stuff to trade."

Sailor looked away, his hand already on the door. "See you in a few."

When they were alone in the Bath, the sun darkening the paper on the windows, Frida kept her eyes on the plastic tubs. She couldn't look at Cal, absorb that neediness. She knew he wanted her to say something like *Can you believe this?* or *Where the hell are we?* He wanted to be comforted by their camaraderie, but Frida was too zombified to offer him anything of the sort. She would wash herself, and that was all. Cal would follow her lead.

She plunged her hands into water so cold it made her teeth ache and scrubbed her pits until her arms hurt. It felt good to be clean. She stepped into the tub next, to soak her feet.

Cal finished quickly, and she watched as he applied hydro-cortisone to the island of dry skin on his arm and aloe to the back of his neck.

Frida grabbed a razor next. Its hollow nothing-weight took her breath away.

Her skin was so dry, and her leg hair so thick, that she winced as she dragged the razor across her skin. "Does it hurt?" Cal asked, but she didn't reply. She moisturized afterward, rubbing the lotion into her calves and even across the tops of her feet. Her skin looked amazing bare, smooth as a slide. She hadn't seen her legs hairless in years, and she'd missed it.

"I wonder if I can get waxed here," she said.

Cal laughed too hard. He'd hang on to that joke for the next eight hours.

He didn't say anything as she replaced her dirty shirt with the one Sailor had given her: it was powder blue and fit just right.

Sailor knocked on the Bath door then. Time was up.

Sailor was to show them the Land. Frida allowed herself to be led around, but she didn't ask any questions. Not about the Spikes surrounding them on all sides, nor about the various decrepit houses where all these people lived, nor about the barn and garden beyond. The tour probably would have been more in-depth had Frida allowed it to be, but every time Sailor asked if they wanted to see something beyond this strip of real estate—they had two cows, apparently, and a herd of goats and a place where residents could sleep under the stars, should they so choose—Frida shook her head. It'd been a while since they'd seen Micah.

"I'd like to see the barn," Cal said.

Frida shook her head again. "I'd like to see my brother."

"Fine, fine," Sailor said, and led them back to the building where they'd left Micah.

In the late nineteenth century, its glory days, it had been a hotel, or more than that. Sailor explained that on the bottom floor there had once been a restaurant and a meeting room for locals and a small store at the back that had sold grains, bolts of cloth, axes. Now, the restaurant kitchen and dining room were where food for the residents was prepared and served. When it was first built, guests stayed in the hotel rooms on the three upper floors, usually for a few nights, but sometimes longer if they were waiting for permanent lodging. It was all in the pamphlet, Sailor said. People used to come to view the decrepit buildings, and the Hotel was one of the main attractions.

"A couple of years before the town closed to tourists," Sailor explained, "money was poured into rehabbing the building, and the Church, too. They obviously ran out of funds before they could finish. But, still, neither is collapsing anytime soon."

"Where is everyone?" Cal asked.

It was a good question. Frida had gotten so used to being isolated, she had barely noticed that all the people who had crowded the main street just an hour before were now gone.

She looked up; on the second floor of a building, a woman was watching them. When they caught eyes, the woman ducked out of sight.

"Micah told them to make themselves scarce," Sailor said. "Until later."

"'Later'?" Cal asked.

They stepped up to the porch of the Hotel, and Frida could tell that Cal was stalling. He wanted to get as much out of Sailor as he could before the others returned. This kid was a talker.

"So where's August?" Cal asked.

Sailor smiled and put his hand on the door. It looked solid, obviously part of the renovation. "He's on another trip."

Cal nodded, as if he expected this. "When will he be back?"

142

Frida didn't think Sailor would fall for it, but he was as care-free as those gophers Cal had planned to capture in his traps.

"In a day or two. Don't worry, you'll see him soon. He likes to relax between trips."

So August lived on the Land. He was one of them. When Frida had told him about her brother, August had known it was Micah she was talking about.

"Let's go inside," Sailor said.

On their way in, Frida tried to catch Cal's eye, but for once he wasn't looking at her. That, or he was pretending not to care about what he'd just learned.

10

He had expected the Hotel to be dark inside, perhaps because the tall windows flanking the front door were draped in heavy curtains, to keep out the heat, he guessed. But he hadn't accounted for the windows on the other side of the building. These had to be uncovered because the room they stepped into wasn't bleak and cavernous but high ceilinged and striped with shafts of dusty light. This was a kind of lobby, though if there had ever been a desk for checking in guests, it had been removed long ago. Perhaps it had been disassembled for a Spike.

Aside from a few worn chairs and an empty bookshelf, the room was bare, its hardwood floor creaky and pocked. A carpeted staircase led to the upper floors, and on the landing a circular stained-glass window tinted the light green.

"Are you guys keeping this as an old-timey hotel?" Cal asked. "It still looks like one."

The ceiling above them groaned. There had to be people upstairs.

Sailor shook his head. "There isn't a concierge or anything." He nodded toward the light at the back of the building, through a wide hallway. "Micah told me to take you into the dining room. I'm sure you guys are hungry."

Cal was, and said so. Frida shrugged, her spine straight, her eyes bright as swimming fish. He couldn't blame her for her shock, but it was starting to unnerve him.

Circular wooden tables filled the dining room, as did a variety of mismatched chairs: a delicate midcentury modern thing that resembled an insect, a few cheap metal ones that could be folded and stacked, even a leather recliner. A long, rectangular table had been pushed along one wall, probably for setting down vats of food for the dining guests. The room's built-in shelves were crowded with trays of silverware and piles of plates, a few bulky towers of bowls, a congregation of motley glasses and mugs, and the occasional Mason jar.

It didn't look like anyone had renovated this room. A papyrus of graffiti covered the peeling wallpaper, and in some places the wainscoting hung slack from mold. One of the large windows had been boarded up with the wood from a patio table: at its center, a perfect circle, for an umbrella, teased like a peephole.

"As I said, the rehab project was never finished," Sailor said. "They ran out of cash."

Cal imagined a velvet rope, cordoning off this back part of the building from tourists.

"It's not that bad, though," Sailor said.

Cal agreed. The other windows were intact and large—was it their light that had penetrated the lobby? Cal realized he was squinting. Even among this disrepair, dining with a sunrise or sunset had to make anyone feel lucky. The view outside was of the land beyond the main street; Micah and these people hadn't settled past the Hotel, and so all Cal could make out was the wild of the woods. He suspected there was a stream or river that way, an easy water source for the residents.

Here, Cal felt so close to the open air, it was like he was standing on the ledge of a train car. He'd done that on the ride from Ohio to Plank; his mom had insisted he take a train to college. The ride was long, but he'd enjoyed the Styrofoam cups of instant coffee and the gnarled man who served them, and he loved to bite the edges of these cups, imprint them with his teeth as the landscape slipped across the window. Ohio was the ugliest, even more so than Nebraska. When he stood on the train ledge, the air had been so strong and rough on his arms, each moment swallowing the tracks beneath him, that he had trembled.

From the windows of the dining room, he could make out more Spikes in the distance. They drew a circle of protection around the Land.

Aside from the boarded window, the others looked to be holding up okay. Only one small pane was broken, and someone had repaired it by taping a Frisbee to the hole. The shoddy fix-it job reminded Cal of Plank. Micah had probably promoted that handyman.

In fact, the whole room reminded Cal of Plank's dining room; it had the same buffet setup, the same disregard for aesthetics. This was probably what Plank looked like now. Had a group of settlers moved in since the school had shut down? He didn't want that. If it couldn't remain as it was, he wanted it to die.

A swinging door at the other end of the room opened, and Peter walked in with an olive-skinned woman who wore men's boxer shorts over leggings. She held a large pot that appeared heavy by the way she flexed her arm muscles. Her tank top was tight against her ribs, and Cal tried to ignore the hard mounds of her nipples. Frida had told him that she and her friends used to put raisins in their bras so that boys would think them perpetually turned on, or cold, their skin brailled with goose bumps. He had never been sure if he should believe her, but now it seemed likely that Frida at fourteen would have wanted to look like this woman.

"This is Fatima," Peter said. Fatima nodded a hello. When she lifted the pot onto the buffet table, he saw that her armpits were unshaved. He wondered if only certain residents were allowed to use the razors in the Bath or if Fatima simply chose not to.

Cal knew that Frida was looking at him, looking at Fatima.

"Where's Micah?" she asked.

"Here I am," Micah said.

He was coming through the swinging door—the kitchen was back there, obviously. Without slowing down, he grabbed a pile of bowls and called to Sailor to get the spoons. Frida had stopped looking at Cal, of course. Her eyes didn't leave her brother, and Cal was nervous she'd fall into an even deeper fog. He led her to the table Micah had set the bowls on and pushed her into a chair that must have once belonged to a 1950s Formica dining set: its seat and back were made of pink vinyl, slit in more than one place, its legs, curved metal. It squeaked with the weight of her.

"I'll get you food," Cal said.

Micah nodded. "She needs to eat."

During the meal, Cal, like Frida, couldn't look away from Micah. Cal's mind had accepted that his old friend was alive, but the specific, distinct reality of Micah was almost too much to bear: the snorted laugh that Cal had somehow forgotten, the way he affected a yawn to fill a pause in the conversation, how he held his bowl of soup to his face to slurp its dregs. It was as disquieting as déjà vu; Cal had been here before, but he *couldn't* have been.

They were eating bowls of bean soup. Cal's spoon was made of silver, and with each bite he tasted the bitter metal. Frida had been given a plastic spork, but she was barely eating. The group had fallen silent, and Cal felt, again, the barrage of questions pushing at him from within.

"I don't understand, Micah," he said finally. "Didn't you blow yourself up?"

"Ah," Fatima said. "No wonder your sister looks so ill."

"We thought he was dead," Cal said to her.

She raised an eyebrow. "That was foolish of you."

Micah held up a hand. "No, it wasn't." That yawn again. "We were very, very good."

"We—as in the Group?" Cal asked.

Peter stood. "I'm clearing these bowls now." He looked to Fatima, then Sailor. "Help me."

When they were gone, Micah leaned forward. "California, you're freaking out my friends."

"Please call me Cal."

"Cal. Sorry. I'll explain everything in time."

"I think right now would be good."

Frida bit her lip. She was watching them like they were in a soap opera.

"Frida," Micah said, turning to her. "I love you. You know that, right?"

She smiled, all at once coming to life.

"Why don't we go see your room," Micah said.

The upper floors weren't as derelict as downstairs, but they hadn't been remodeled, either. Not professionally, at least.

"We've spent some time making things comfortable up here," Micah said as they headed upstairs. The walls were painted, if a bit sloppily. Cal could make out burn marks on the ceiling from the flames of gas lanterns long ago.

Their room was on the third floor near the end of the hallway. It was a small room, with a small glassless window that was covered with a piece of cheesecloth. The thin fabric had been stapled to the wall.

Their bags were waiting for them on the floor, but they fell slack, as if someone had removed most of the belongings inside.

"Did you take our clothes?" Cal asked Micah.

Micah smiled. "They're probably just being washed. Don't freak out."

"My jeans were in there."

"You really care that much about *jeans?*" Micah said. "I wouldn't have taken you for such a fashionista, Cal. Besides, I said, you'll get everything back."

"When?" He was about to say more, when Frida grabbed his hand and nodded at the room, as if to offer it to him. But he couldn't wear a bedroom.

There was a collapsible camping table made of nylon and metal, and a bed with a wrought-iron headboard. Cal couldn't tell if the sharp, itchy mattress of hay was from the nineteenth century or if it was merely supposed to appear that way.

"You'll get used to it," Micah said, when Cal sat down on it.

"People always say, 'You'll get used to it,'" Frida said suddenly. "It's not really true."

Micah grinned. "That's just because you're spoiled."

Frida laughed.

Cal turned to Micah. "You're the one with clean fingernails."

They both looked at him like he'd gone too far.

"No need to get sensitive," Micah said. "I was only poking fun."

"So was I," Cal said.

"Boys," Frida said. But then she was quiet again, and Cal couldn't help but be annoyed.

"Is this room usually empty?" he asked. "Is that why there's nothing in it?"

Micah shook his head. "We're not too big on personal effects here." He explained that this was Fatima's room.

"Where will she sleep?" Cal asked.

"Probably with Peter."

"Is Peter her—husband?"

Micah burst out laughing. "Can't be married without a government, California. If you're asking if they're a couple, then, yeah. I guess. I don't ask. But they haven't jumped over a broom or anything." He squinted at Cal. "Did you guys? Are you?"

Frida nodded.

"A year or so after you . . . left," Cal said.

Micah stroked his beard. "I see. So, what? You're Frida Friedman now?"

She laughed. "God, no. Can you imagine? The choice to keep Ellis was a no-brainer."

"I didn't realize that kind of thing—marriage—still happened," Micah said. "And so young, too. That must've been quite a surprise to everyone."

"Not really," Cal said. "My mom always said I was an old man trapped in a kid's body. I was only twenty-three when we did it, but damn if it didn't feel like a decade had passed since I left for Plank." He tried to smile. "People said it seemed right for us. Besides, it's not like we had a wedding."

"No? What a shame," Micah said.

"Your parents gave us their rings," Cal said, "but we sold them, for money to get out."

"Good for you," Micah said.

No one spoke for a moment.

"Can you tell us how it is that you're here?" Cal asked. "Please."

Micah scrunched his lips together. "It was a prank, Cal. An elaborate prank."

"But someone blew up. People died."

"What made you so certain it was me?"

"The Group claimed it," Cal said.

In good old terrorist fashion, the Group had issued a statement five hours after the bombing. They'd released a photo of Micah and some gibberish about taking back public spaces for the common citizen. The police had confirmed his identity; there had been DNA evidence.

"Did we ever claim anything before that?" Micah asked.

"Come on, you'd been posting your stunts online for years."

"But we never officially owned up to anything violent."

"That doesn't mean you weren't responsible."

Micah sighed. "Look, I'll tell you this much." He took off his hat. Cal saw nothing until Micah bent forward. There it was on the crown of his head, a dime-sized bald spot in his mess of hair, pinkish and rough edged

"Did you get that in the bombing?"

"I got it a few hours before. Hurt like a motherfucker. That, and we drew a bit of my blood. Put it on a vest we planted there."

Frida spoke up. "Someone scalped you?"

Micah put his hat back on his head. "Just a patch."

Cal shook his head. "That's absurd."

"Bingo."

"Did anyone blow themselves up?"

"Yes, just not me."

"Didn't they want the credit?"

"Credit didn't matter to him. He was dedicated to the cause." Micah sighed. "It was important that everyone think I'd done it. It was proof of my commitment to the Group."

"Except it wasn't you."

Micah nodded. "What can I say? I was needed elsewhere.

Anyway, the bombing drove away anyone who wasn't serious about the Group's cause, and we recruited a bunch of new members."

Cal nodded. "No more tap dancers, right?"

Micah laughed. "Exactly."

"People died," Cal said. "Including some brainwashed maniac. He died anonymously for your cause." When Micah didn't reply, Cal asked, "Why did you do it? Was it really to protest that awful mall?" He looked at Frida, but she didn't even seem like she was listening. He wanted to pour a bucket of water over her to get her to wake up.

"It could have very well been a fancy gym," Micah said, "if they hadn't all gone out of business by then." He paused. "If you remember, Hollywood and Highland was still attracting business from Community members. They'd landed an exclusive deal with Calabasas, and another with Malibu was in the works. Those Communities agreed not to shop anywhere else, and in exchange, their members got special discounts. Not to mention the secure shuttle rides."

"Other people went there, too, though," Cal said. "Not just Community members."

"But how long would that last?" Micah asked. "There were plans to cordon off certain sections of the mall."

Cal shook his head. "So what? The people who worked at the mall lived in L.A.! What about them? And the money Calabasas and Malibu spent went back to us, into the city. Didn't you stop to think about that? Everything went downhill after what you did. Hollywood and Highland closed, and the Communities went ahead and built their own malls. Good work."

"They're called shopping plazas," Micah said. "But yeah, I know."

"I guess that means you didn't totally fall off the edge of civilization," Cal said. Frida looked up and nodded, as if there was something to agree to.

"It wasn't a perfect plan," Micah said, "I'll be the first to admit that. But it invigorated the Group like nothing else would have, and the encampment was stronger after that, as you're probably aware." He fought a grin. "And then there were all those copycats, which I honestly didn't predict."

Cal felt himself sneer.

"Everyone's meeting at the Church before dinner," Micah said as he turned to leave, "so you'll get to see them. You'll sit in a pew, but don't worry, that's where the religious stuff ends." He smiled. "This isn't a bunch of believers."

"You do look like a cult leader with that beard," Cal said. He forced himself to smile, to show that he, too, could take it easy, let go a little.

"I wish I could cut it, man," Micah said. "But every time I try, I get a weird rash."

"Creams," Frida said suddenly.

"Yeah, use some of the stuff in the Bath," Cal said.

"Maybe," Micah replied. Perhaps because he could feel more questions coming on, he stepped back. "I have some things to take care of. You'll need to wait here in your room for a few hours until I'm done." He nodded at the closet door. "There's a bedpan in there, should nature call. We've also got latrines, but I'd rather you didn't wander off until everyone on the Land's been brought up to speed about you." He paused. "I'm not having anyone stand guard outside your room or anything, but please don't make me regret that."

Frida nodded again, as if she didn't mind being a hostage. Cal didn't say anything.

*

Micah closed their door, and once they had privacy, Frida took off her shoes and moved to the floor. She began stretching, going through the basic vocabulary of the few yoga classes she'd taken in L.A.: downward dog, cat-cow, child's pose. Her breath became slow and throaty, almost mournful. Cal leaned back in the bed and watched her. The metal knobs of the headboard dug into his skull, but he let them.

Frida had turned to stretching as a way to counteract anxiety; after weed became too expensive, this was the only thing that worked. Cal could remind her to breathe all the time, but such exhortations only worked if she was willing. He loved that she stretched, that these poses relaxed her nerves. She fought the body with the body.

As she moved through the poses, he thought about what Sailor had told them. August lived here. He was a member of this tribe, not some lone trader. Frida had looked so stunned by the news that Cal had forced himself to look away. If he hadn't, he might have done something rash: punched Sailor in the mouth, or laughed at Frida like a maniac, or even fallen to his knees.

Now Cal wanted to ask her how Sailor had known Frida's name. How had he known she was Micah's—*Mikey's*—sister? Maybe Micah had told the Land all about his older sister, and August had finally put it together who Frida was. But Micah was never one to blabber about his personal life, especially not after he'd joined the Group, and August didn't seem like the type, either.

On their drive away from L.A., Frida and Cal had agreed that they wouldn't tell anyone about her brother's ties to the Group. No one had to know her brother had died or that she'd even had a brother. Cal said it was for safety; Frida had said it was for solace.

It had been easy to follow this rule with the Millers, who themselves acted as if there were only the present and a glorious, pure future. If Frida had tried to confide in Sandy, Sandy would have certainly shut her up.

But this had nothing to do with the Millers. Frida must have talked to August about her brother—he didn't explicitly trade for secrets, but that was what he was after all along, wasn't it? It was obvious, Cal decided: August had returned from trading with Frida to tell Micah that his sister was just a couple of days away. That his sister was not only nearby, but pregnant.

If Cal and Frida hadn't come to the Land, Micah might have come to them.

"This bed is awful. We can sleep on the floor tonight," Cal said.

Frida was lying on her back with her arms at her sides, her hands loose. Corpse pose.

Cal slid off the bed and peered into his backpack. His jeans and his shirt had been removed, as had the flashlight and the sleeping bag. All that remained were his empty canteen, his sweatshirt, and a pair of socks.

"I can't believe they took our stuff. Why would they do that?"

Either these people were playing mind games, and they wanted Cal and Frida to feel needy and vulnerable, or they were just used to grabbing whatever they needed, whenever they needed it. Maybe ownership meant nothing on the Land, and any old possession could be taken from you, at any moment.

"I doubt my jeans are getting washed," he said. He turned back to Frida, who hadn't moved from the floor. "You should check your stuff, too, to see what's left."

Frida didn't reply.

"You okay?" he asked.

"Micah is alive."

"I know," Cal said. He left his bag and sat down on the floor next to her. Her eyes were closed, but she didn't flinch when he laid a hand on her stomach. It was warm, and her body moved with her breath.

"Kiss me," she said, and he leaned down and did.

She pulled him on top of her and began kissing him more intensely, her hands on his back, crawling under his shirt. She wanted him. He felt how alive she was beneath his body, even after the shock. Or because of it. Cal pulled away.

"When do you think he'll bring up your pregnancy?"

Frida didn't say anything for a few seconds, and then: "Probably never."

"Huh." In a way, it made sense. Micah didn't seem like the uncle type.

He tried to kiss her again, but she moved away.

By the time Micah came to pick them up that evening, Cal was hungry again. Frida wasn't complaining, but he wanted her to be eating as much as possible. Didn't morning sickness strike in the first trimester? If it did, she'd need all the calories she could get while she was well enough to get them.

Micah had changed his clothes. He wore a button-down shirt, the sleeves rolled up.

"Sunday's finest?" Cal asked.

Micah laughed. "Man, I didn't realize how much I missed you."

Cal couldn't help but laugh, too. Maybe this would be okay. Maybe all he'd needed was some rest and a little more information.

"Where's my stuff?" he asked.

Micah raised an eyebrow. "I told you. I'm sure Dave just took them to get washed or something."

"And the flashlight?"

Micah nodded. "Oh, don't worry, he probably didn't put *that* in the wash." When Cal didn't laugh, Micah said, "I'll look into it."

"Please do," Cal replied. "And can we get Frida some jerky or something?"

Micah frowned and looked at his sister. "Are you hungry?"

Frida said she wasn't.

"We won't be long," he said to Cal.

They peed in the bushes behind the Hotel. "It's not prohibited, but not really encouraged, either," Micah explained afterward, as they walked to the Church.

People were coming out of the buildings now. They were done making themselves scarce, Cal supposed. He recognized a lot of them, not just Fatima and Peter and Dave but the heavy-set man who had given him a dirty look earlier and the blond-dreads guy.

Others were strangers, and their newness made him feel strange. He wanted to start giggling, but he held back and tried not to look at one face too long. There was a woman with long black hair and an open, round face; the guy she was with was tall and very pale, with a slump to his shoulders that made Cal's neck hurt just looking at him. He was younger than the woman, probably by about ten years. Right behind them was an Asian man, not much older than Sailor, who walked with a slight limp and smiled at Cal and Frida when they passed. He was the exception; most of the people tried not to look as Micah led them onto the main road to join the procession.

The sun was going down, and it was getting chilly. Cal had forgotten his sweatshirt in the room—he could see it balled like a possum at the bottom of his backpack—and now he crossed his arms to keep in the heat.

"Cold?" Micah asked. "By the time we're done, the Church will be sweltering."

It sounded to Cal like a threat, but he couldn't figure out how exactly.

The outside of the Church had been lit up with torches, and they lined the walls on either side of the entrance. As he got closer, Cal saw that they weren't handmade, as he'd imagined, but the cheap tiki ones that people used to purchase for their outdoor parties, their ersatz luaus, so many years ago. He hadn't seen the things in a long time. How was it that the Land had any? Cal tucked the question into his mental file of things to ask Micah.

He peered at the building before him. Except for two small windows near the roof, the Church looked like a big, enclosed square.

There had to be two floors, Cal surmised, or an attic. Perhaps it was windowless on the bottom floor so that it could be used year-round, despite the weather, or it could be a place to hide during a tornado or a hailstorm.

He wondered if the double front doors were more secure than the door at the Miller Estate. If a town of people crowded into this church at night, without candles or tiki torches or solar lighting to illuminate the space, would they be able to see one another?

The small windows on the upper floor were, miraculously, intact. Miraculous, indeed. More like unbelievable, Cal thought. Micah had probably had them replaced in the last year or two, the panes coming from the same source as the razors, the tiki torches, and his Polo shirt.

It looked like the outside of the Church had been painted

professionally, and though its surface was chipped and stripped in sections, patched here and there with mismatched colors, it certainly wasn't a nineteenth-century undertaking. Roaming bands of settlers must have defaced the building in the years since the town had closed, and the Land's solution to this vandalism had been to paint over it; here and there, the Church's stark white surface was marred by squares of beige, eggshell, and even pistachio green. It looked a little sloppy, but Cal liked the bedraggled quality it lent the building. Now it was less somber church and more local high school. Besides, the steeple rising into the sky looked as unscathed as it must have been two centuries earlier; the barbed wire around its body merely confirmed its glory, made its pierce into the heavens more powerful.

A sign at the front welcomed visitors. In the fading light, Cal couldn't read the text clearly, but it was obviously for tourists, most likely describing the religious life of the town before it had been abandoned. Micah pulled him along, said, "You'll have plenty of time to read that later." With his thumb and index finger, he flicked the air between him and the sign. "Besides, those moron amusement-park developers didn't care about this place's history. They fucked this building up. It doesn't look as old as it should, as it *is*."

Inside, it was so bright that Cal's eyes watered. He hadn't seen light like this since leaving L.A., and even then, electricity had been scarce. He thought of gas stations in the middle of the night; of his mother's desk lamp, which she'd fed with high-wattage lightbulbs, despite the threat of fire; of the streetlight that used to burn into his bedroom window unless he secured the curtain closed. All that, years ago.

"I need sunglasses," he said to Frida. "I feel like I'm stepping on-stage."

In each corner, an industrial light, the kind used on construction sites, was connected to a car battery. How wasteful it seemed. The lights emitted a terrible droning buzz. Such a noise would've been normal just five years ago, but now it struck Cal as insidious—unbearable, certainly, should they have to sit near one. He looked at Frida, who had placed her hand on her abdomen, as if to palm their child's ears. Cal's stomach dropped. The sounds of technology, the insistent whirs and hums and sighs of motors, computers, lights, clocks, cooling and heating systems, masked an entire, secretive universe, a world beneath the world. Their child would be, could be, *should* be, a creature capable of discerning the smallest shifts out of silence. Like a woodland creature, ears pricked to the slightest movement miles off, he would truly be able to listen. *Listen.* He imagined Micah, or his suicidal dupe, saying that word, and his stomach dropped farther.

"Our meetings can't run too long for this reason," Micah said, and nodded to the lights. "We don't want to waste our resources."

"Resources," Cal repeated. He didn't bother asking why they just didn't use candles for the meeting. Clearly, these lights played into Micah's theatrical streak.

"Follow me," Micah said.

People were already packed into the pews that began just a couple of feet from the entrance and continued in orderly rows toward the front of the large square room. The walls were made of plaster and blank, without iconography. The ceiling was high. Micah had said that they weren't a religious group, and Cal was relieved.

The second floor was most likely accessed through the unassuming door behind the raised stage. There was a podium on that stage—a pulpit? Was that the correct term? Frida would probably wonder the same thing, but the difference was she wouldn't be embarrassed that she didn't know for sure. Both of them had been raised heathens; that had been Frida's father's word, said with a snobbish little guffaw—but it was true. Cal's parents weren't believers, so neither was he. He occasionally prayed, as he had done before their journey here, but it was a pitiful begging to no one in particular. He didn't see the point of worship.

This was probably only the fourth or fifth religious establishment Cal had ever set foot in, including the tiny storefront church down the block from their apartment in L.A. That place had low ceilings, and three rows of plastic patio chairs had faced an altar covered in a disposable tablecloth. There were spelling errors in the literature (even in the Spanish text), and about as much atmosphere as a Laundromat. But the people there had been so taken with the Lord. *Jesús es Dios,* they told him, clutching their Bibles, their babies. An old woman had led him in; he'd been on his way home from work, drunk from the home-made cider a coworker had brought, and he'd thought, Why not. *"Pues,"* he'd said to the woman, *Well,* and walked inside. Those churchgoers had spoken of end times, which would have made Cal uncomfortable when he was in Cleveland or at Plank, but by the time he lived in L.A. such talk was commonplace. Before CNN had gone dark, he'd heard the phrase tossed around by most of the pundits. *Pundits, pulpits.* What was the etymology of these two words?

Frida brushed against him as they moved down the line of pews. The aisle was narrow, and yet they walked side by side, as if in a wedding procession. Not that this building had seen one of those in a while. Even so, this was just a church, in the end, clean and unadorned. The Land had kept it as uninteresting as the bedroom he and Frida were staying in.

"Where's Sailor?" he asked as they got to the first pew, left empty.

"At the lookout Tower," Micah replied. "On duty." He ushered them to a seat.

"Is that his punishment?" Cal asked. He wanted to add *for spilling the beans,* but Micah's confused look made him shake his head. "Never mind."

Cal slid into the pew first, then Frida, and then Micah next to her. Cal faced forward because he didn't want to seem too eager to sum up this group of people. They believed in containment, which probably meant they were skeptical of curiosity as well.

The women behind him were talking in hushed voices. He leaned back to listen.

"If you want to keep that shirt, you'll need to retrieve it from the line while it's still damp, before anyone else takes it for themselves. Not everyone here is as crazy about wearing the same clothing all the time as you are, they like to mix it up. And then you get upset."

"I know, I know." This woman gave a little laugh. "I shouldn't care."

"But you do. We all have those things. We can't help it, even Mikey admits that."

"True. I should just wash it already. It stinks!"

They both laughed and fell silent.

Cal placed his hand on Frida's thigh and kept his gaze ahead.

Micah had been right: this couldn't have been how the Church looked originally. Surely, the ghost town's stage would have been built of rough wood planks like everything else, or maybe bricks. In its place, the developers had built a smooth sanded stage, with a piping of metal around its soft edges. The door behind the stage matched.

Who was waiting behind the door? What was on the second floor?

A few people ventured to the front of the room to see Micah. One asked about something called Morning Labor, and another came to apologize for drinking the milk Micah had requested. "You never came for it. It was on its way to curd," the man said, and laughed, trying to catch Cal's eye and, when that didn't work, Frida's.

Micah sent away all of these visitors with a curt nod or a subtle shake of the head.

When he put his arm across Frida's front and said to a visitor, "We'll begin the meeting shortly," Cal realized there was no one behind the stage door. The meeting would start when Micah started it. He was the televangelist here.

And just like that, the voices in the pews behind them faded away. Cal heard someone close the front doors, and a man's baritone groan about the heat that would descend soon enough. The doors were reopened, and someone else, a woman, complained about the bugs that would soon be in the Church, attracted to the brightness. No one listened to her. Micah hoisted himself to standing.

He walked up the two steps leading to the stage and stood behind the pulpit. There had once been a microphone up there, no doubt, but now whoever wanted to keep the congregation's attention simply had to project. But it wasn't a problem, Micah's voice was so loud, Cal leaned back.

"Peter was supposed to run tonight's meeting, but it looks like you're stuck with me."

Cal had expected a more eloquent welcome from his brother-in-law, some comment on the evening, the day's momentous events, but Micah acted as if nothing new had happened that day, as if he hadn't in fact come back to life for his own sister. The lack of preamble should have eradicated Micah's charm, but the people's silence proved that they respected him, were magnetized by his presence alone.

Or maybe this was just normal. Perhaps the Land didn't require niceties, fancy speeches. They'd stripped away the fake and dangerous veneer of modern culture, the one Cal himself had been eager to leave, in order to live freely. Micah was just being himself up there, and people were listening, not in the name of etiquette, but because it didn't occur to them not to.

"I know we need to discuss Morning Labor and the issue with people missing their shifts, but . . ." Here Micah smiled, and a few people at the back of the Church laughed. Cal thought he heard someone stamping their feet.

"You all learned today that I have a sister and that she's been just a few miles off for the past two years." He paused. "As you probably already know, she's here now. She's come to the Land."

Frida was leaning forward in the pew, her hands shaped into a steeple like the one above them. She reminded Cal of a high school basketball player, watching the game from the bench, hoping to be called in. Micah said her name, and her hands fell. She turned around to take in the crowd. Cal kept his eyes on Micah.

"Frida's with an old friend of mine. His name is Cal. Short for Calvin, but call him Cal. Everyone does."

So Cal wasn't Frida's husband, or even her partner, or her boyfriend. He was Micah's pal.

"Frida and Cal," Micah said, lowering his voice to address them. "Stand up, so that everyone can see you."

Frida was on her feet before Cal could even compute the request. She pulled him up to join her, and Cal finally took in the congregation. The Church was crowded with people, more than just the ones he'd seen earlier, probably fifty or sixty. They were sitting in the pews or standing by the open door. Those sitting by the lights were already soaked in sweat; Cal himself could feel the wetness under his arms and at his forehead.

Some of the people were grinning at him; others were nodding solemnly. He caught Fatima's eye, and she raised both eyebrows in a goofy way. When he tried to make eye contact with Peter, he just looked through him. Cal shrugged and looked away.

No one was elderly. No one was very young.

It hit him all at once. There were no children. Not one.

"Where are the kids?" he whispered to Frida.

She sat down as if she hadn't heard him.

"I know this situation isn't to be taken lightly," Micah said. "I don't expect any favors." He rubbed his hands together. "We will bring their presence here to a vote. Believe me, we will."

If the Land voted against their presence, Cal wondered, would he and Frida cease to exist?

"But, in the meantime," Micah continued, "I'd like to have them stay here, just until a decision is reached."

Now that Cal had already seen the crowd, he couldn't help but turn around to witness their reactions to Micah's request. He half expected someone, Dave, maybe, or Peter, to stand up with a pitchfork and demand the outsiders leave. It would make sense; Frida and Cal's presence expanded a community that wanted to remain the same size.

He wanted to stand up and say, *Do what you want. We're happy to leave.* He wouldn't, of course. And, anyway, Frida didn't agree with him.

"Is that all right with everyone?" Micah said.

Cal had to grab on to the back of the pew when the women sitting behind him held up their fists in response to Micah's questions. They moved them back and forth, as if their hands were hinged, as if they were knocking on invisible doors.

Just like at Plank. The way they'd expressed approval, whether they were cleaning horse stalls or discussing Roland Barthes in seminar. Why was the Land mimicking that knock? Micah had clearly taught them that signal of approval; or had some of them already known it?

He began feverishly looking around the room at the other men's faces. Did he recognize anyone? Except for Micah, he hadn't seen any of his classmates since the day he'd left Plank. What if they were here, and he could talk to them again? What did that mean? A ghost town that lived up to its name.

But he didn't see any Plankers. Strangers stared back at him, and his new startling hope flew, startled, out of him.

A couple of people weren't knocking, he realized. They held up their index and middle fingers—Plank's signal for disagreement. He turned back to Micah.

"Please come speak to me individually afterward if you don't agree," he said.

To Cal's surprise, none of the dissenters stood up to protest. This was a civilized bunch. That, or Micah had power over them. Maybe here on the Land, democracy was merely dress up, merely a dance. They had the stage lights for it.

"I promise," Micah said, "we'll vote on this very soon." He smiled. "Until then, I urge you to get to know Frida and Cal. I'll be putting them on the Labor schedule. They've been living on their own out here, so they're strong, and resourceful."

Cal realized Micah was serious. They'd been welcomed, albeit temporarily, to this place, just as Frida had hoped, and he had feared. They'd be put to work, which was clearly important here; Micah was already discussing the Morning Labor controversy. Cal and Frida would become part of this world. Frida would be pleased. Cal wasn't sure what he felt. They had squash back home that would soon need picking and a bed more comfortable than the sack of straw Micah had given them and a house that fit the two of them, a third when the time came. If Frida ate enough, their baby would be healthy. August could bring them special goods, if needed. Clearly, the Land had access.

Maybe that's what troubled him. This wasn't a ghost town at the edge of the world. They were connected to something larger.

After the meeting, they ate more of the same bean soup in the dining room with about thirty others. The Land dined in two shifts, Micah explained, and some, if they were too hungry to wait, prepared simpler meals in their own houses. The room was lit with candles and a single solar lamp, far brighter than the ones Cal and Frida used.

Once they were seated, Cal asked Micah, "So does everyone here know about Plank?"

Micah shrugged. "About as many as in the real world. Which is to say not many."

"But the fist knocking . . ."

"You noticed?"

Micah's pretend ignorance made Cal throw down his spoon. "Don't act like you don't owe us answers," he said.

Micah took another mouthful of soup. His eyes shot left and right as he did so, and Cal could tell he was trying to gauge the tension, the interest, in the room. He didn't want the visitors to cause a scene. Cal was willing to do a lot more, if it got Micah talking.

"Mikey?" Cal said. "Tell us what the hell is going on here."

Micah didn't answer, just sat there, silent. Cal made to stand up, but then Frida's hand was on his.

"Later," she said. She nodded to her bowl of soup, and to his. "Let's wait, okay?"

"Yes, she's right," Micah said. "It's dinner, and we should just enjoy it."

Frida brought her spoon to her mouth and ate. He hated how content she seemed, that happiness she couldn't conceal.

The next morning, when they were alone, Frida changing into the clothes Fatima had dropped off for her the night before, Cal tried to talk about the turkey baster.

"Why didn't you tell me we had that?"

She shrugged, but he waited. She couldn't play mute for good.

"Tell me. Frida?"

"It was fun to have a secret," she said. "And I didn't want you to take it from me."

"Why would I take it from you?"

"To use."

"I don't mind if you give it to Micah," he replied.

"Don't be jealous."

He grunted. "I'll try not to be."

She didn't reply.

"Just think about it, babe. These people have batteries and razors—sharp ones. Last night Sailor was carrying one of those glow sticks they used to use at raves, lighting his way. They're not desperate for things, not like we are."

"I'm glad you can finally admit it. We're desperate."

"You think the baster will impress him, but it won't."

She looked at him like he was the lowest human being.

"I'm not interested in impressing him."

She straightened her shirt, which fit her loosely. He was reminded of the first time she'd worn Sandy's clothes. He'd never told her how much it upset him.

"Did you hear what I said last night? In the Church?"

She looked up, waiting.

"There aren't any kids here. No babies, either."

She didn't answer, but her face said: *Don't ruin this for me.*

"It's not as if they were in day care, Frida."

"What do you want me to say? You think we're in one of those fantasy novels you read when you were little because you didn't have any friends?"

"No, Frida, please. I'm not talking about some impossible future world. I mean something perfectly logical . . ."

"Like what? That the tribe can't get pregnant? Like the world is seriously ending? Please."

"It just worries me."

"Stop worrying, okay? Just for a while."

He didn't reply.

In the next moment, Micah was calling their names from the hallway.

Freeeeda. Californeeea.

"Your leader is calling," Cal said, and stepped aside so she could pass.

II

Morning Labor wasn't as bad as Frida had expected. The name itself had scared her, but in reality, it was just a list of chores that the Land members had to complete before noon. These positions were assigned by a committee, and they rotated monthly unless someone was particularly skilled at a task and wished to continue doing it permanently. She and Cal were supposed to choose from a number of assignments: kitchen, garden, construction, butchery, security, animals, laundry, or housekeeping. Per regulation, they were told they could not pick the same job. Couples separated before noon to encourage socialization and independence.

When Sailor had explained the system to them after dinner the night before, she'd thought immediately of Plank. Of the jobs Micah had complained about in his letters, and of Cal's stories, told in such detail that she could trick herself into believing that she'd gone to the school herself. The Plankers, she knew, had alternated positions in the same way they did on the Land, and they took them just as seriously.

As Sailor continued, Cal had leaned over and whispered to her, "This is exactly like—"

"I know," Frida said. It would have been unkind to pretend she didn't notice the similarities.

Now, their first morning here, Micah stood with Sailor in the hallway. They were both wearing thick sweaters, and Micah had on a beanie.

"Good morning," her brother said.

"Labor's about to begin," Sailor said. "You should get there early."

"Have you chosen an assignment yet?" Micah asked.

Frida picked kitchen because of her baking experience. "And Cal . . ."

He was just stepping out of the bedroom, not even trying to hide his scowl.

"What'll it be?" Micah asked him.

"Whatever," Cal said. And then, after a moment, he added, "Security."

Micah shook his head. "You aren't familiar enough with the Land for that position."

"How about construction?" Sailor asked, looking to Micah. Her brother nodded.

Cal shrugged. "Whatever," he said again.

Micah smiled, as if he didn't notice the attitude. "Perfect. I'll have Sailor lead you guys to your assignments."

He nodded once more to Sailor and, without even a wave, left them to it.

Frida walked into the kitchen and realized she'd been imagining the one at Canter's, which could serve two hundred diners if needed. They'd certainly baked that much bread and pastry each day. This kitchen, still dank in the gray morning light, was much smaller, and of course it wasn't outfitted with industrial ovens and dishwashers. For washing dishes, there was a rusty trough next to a back door; the trough was presumably filled with water from outside, and the buckets waited on the floor nearby for such a task. Across the kitchen, a large woodburning stove stunk up the room. Pots and pans hung from the ceiling, and in the center of the room ran a long banquet table; for chopping and prepping the food, Frida guessed. One end was crowded with cooking tools: cutting boards, large mixing bowls, knives, slotted spoons.

As Frida walked in, a middle-aged woman with a gray streak in her hair was ascending from what had to be the root cellar. In one hand she held a basket of onions, and with her other hand she closed the two wooden doors that stuck open, vertical, from the floor.

She saw Frida and smiled. She was missing a front tooth, and the ones she did have were yellow and uneven.

The woman put the onions on the table. "I'm Anika."

"Frida."

"I know your name," Anika said. She glanced out the window. She was checking the light, Frida realized. "You're early. The others will be here soon."

Frida nodded, unsure of herself. She remembered what Micah had said about her and Cal during the Church meeting: that they were strong and resourceful. She would have to live up to that promise.

Before she could ask if Anika wanted her to start on anything, the others on the shift arrived. There were seven of them in all, including Fatima, who was wearing the same outfit she'd worn since Frida's arrival. It looked like she'd slept in the boxers.

To Frida's surprise, Fatima came over and gave her a kiss on each cheek.

"Oh ... hi," Frida said.

"Don't look so starstruck," Fatima said. "It's how I greet all of my friends."

"That's the thing," Frida said with a laugh. "I haven't had a friend in a while."

Anika began her instructions. She wanted them to cut

onions, she said, and peel carrots and slice potatoes so thin they were see-through.

"And I mean thin," Anika said, "like skin."

Frida quickly learned that Anika was on permanent duty, and thus in charge.

She was the team leader, Fatima said.

Most of the group was female, except for two guys who were as young as Sailor. They didn't seem quite as naïve or dewy as he was, though if they'd told her they played in the same band, Frida would have believed them: they were scruffy enough, skinny enough. One had a tattoo of a feather on his thumb and asked Frida if she could handle cutting onions. The other went right for the potatoes as if Anika's instruction had inspired him.

When one of the women saw how slowly the boy was working through the first potato, she said, "It's not easy to cut them as thin as we want them." The woman had told Frida her name was Betty. Her hair was a cloud of dark ringlets, and her large brown eyes reminded Frida of a doll's. "We need a mandoline," Betty added.

"Noted," Anika replied from the washing trough.

The boy shot Betty a fierce look, and both of them glanced at Frida. She immediately went back to her onions, which were making her nose run, her eyes water.

Eventually, she stopped resisting and just let the onions do what they would do. Her eyes were stinging so much they seemed to spasm, and the tears ran down her face. But she didn't stop dicing. This knife was sharper than any she and Cal used, and she liked its weight.

Time slipped by. When she finished with the onions, Anika complimented her technique and handed her bulbs of garlic to mince. The guy with the feather tattoo had the same job, and they stood side by side, in silence, pushing the cloves with the flats of their knives so that the skins cracked open. Frida's back began to hurt, the way it used to when she'd been baking all morning, but she didn't even stop to roll her shoulders or hang her neck forward for a little relief, though she saw others doing so, even Anika.

Someone started humming a song. A lullaby. *Rock-a-bye baby, on the treetop.* After the first phrase, a couple of others joined in, including feather-tattoo man, which made his friend shake his head and snort like a pent-up animal.

When the wind blows, the cradle will rock.

Morning Labor. Was that supposed to make her think of having a baby?

No kids, Cal had pointed out. She didn't want to talk, let alone think, about it. Because if these people couldn't have kids, or if they didn't allow them, what would that mean for her own child?

Cal thought Micah knew about the pregnancy or, at the very least, that August did. He had no idea that there was a secret to keep, that she and Cal were the only two people on earth who knew about the tiny human inside of her. There was something beautiful about that kind of secret.

She knew she'd have to tell her brother. That's what Cal would advise. Maybe they did have a urine test here, and she could find out for sure. Oh, please, that was unnecessary. She was almost three weeks late, and that had never happened. She was pregnant; she just knew. Wasn't that how it worked? Before long, smells would turn her stomach, as would certain foods. She might have to carry around a bucket, maybe a barf bag, which the Land no doubt had a box of. She'd need crackers to calm her belly; if they let her bake, she could prepare them as she preferred. By then, there'd be no hiding her malady. Not that being pregnant was an illness or a handicap.

If she was pregnant, she'd raise the baby here, on the Land . . . if she survived the birth.

Stop it, she told herself. *Stop it.* She was enjoying herself, and she didn't want to ruin the morning with anxiety. It felt good to cook like this. So what if she was pregnant? So what if her brother had treated their reunion so casually and didn't seem to want to be alone with her? She was back in a kitchen cooking with others, in a room with windows. She was grateful.

Sometimes, a conversation would begin at one end of the kitchen and, just as quickly, extinguish like a match in the wind. A few would start giggling about something Frida couldn't hear, and then Anika would announce something briskly to the whole group—an encouragement or a technical reminder about how to hold a knife—and the mood would turn serious again.

She liked being part of the routine. It reminded her of the Canter's kitchen at 4:00 a.m. when it was just her and the other bakers and a few prep cooks who would pass behind her warning, *"Por detrás,"* as they balanced cutting boards of sliced tomatoes, their slippery seeds sliding off the edges.

As it had been back then, it was easy to focus on each rote task given to her. She didn't mind that Anika eventually assigned others more complicated work: scaling and deboning fish that a man named Charles had caught in the nearby river and brought in through the back door, for instance, or conferring with Anika over the menu, discussing substitutions and portion sizes. She could tell them later about her skills; for now, she would bend over the table and cut cloves of garlic, one after another: like waking up to a new day, every day. Her fingertips were sticky.

"You can relax a little, you know," Fatima said from behind her, a tray of deboned trout in her hands.

"I'm relaxed," Frida said. "Just communing with my garlic."

"Sailor said you've worked in a kitchen before."

Feather guy looked up. "Oh yeah?" he said.

Anika was across the room and didn't seem to notice them talking.

"I was a baker, at a deli in L.A.," Frida said.

"Bread? Can you do bread?" he asked.

"Of course," she replied.

Fatima explained that they had a small bread operation in place. Their wheat harvest had turned out beautifully the last two years, but no one was really very good at baking.

"The sourdough's bland," Feather Boy said.

Fatima rolled her eyes. "Burke is very hard to please."

Burke shrugged like a dad in a sitcom, and Frida wished he or Fatima would call out to Anika and announce that the new girl, Mikey's sister, was a professional, that she could bake, that she could do bread. She hadn't baked in years, and the idea of using the woodstove made her nervous, but already she could smell the dough, hear it rising (it did have a sound, she swore it did, it was like a gathering of energy). The leave-it-alone mantra, coupled with that urge to knead, and the way the loaves felt just baked, warm as breathing bodies.

If these people tasted her bread, they would definitely allow her to stay. If the fact that she was Micah's sister didn't give her special status, then her talents would.

She'd been distracted by this little fantasy when her knife sliced into her finger. "Ow," she said without wanting to, and brought her finger to her mouth.

Anika was next to her in seconds. "You all right?" she asked.

Frida nodded. She felt like a little girl, sucking on a lollipop. "Just a little cut," she said. "I'll live." She showed Anika the wound—a torn flap of skin. "Thin as those potatoes," she said, but Anika didn't laugh.

Blood seeped into the cut, a diagonal red line, and Anika turned away.

"Do you have a rag or something?" Frida asked. She remembered the tin of Band-Aids the Millers had brought her, so long ago. She'd only used a couple of them.

"You'd better just wash it out," Anika said, not meeting her gaze. "And hurry."

Burke continued cutting his garlic as if he hadn't even noticed what had happened. Fatima was now on the other side of the kitchen, doing something with her pile of fish. Everyone seemed too focused on their work, like they were acting in the same terrible play. It was as if they were embarrassed for her.

Frida put her finger back into her mouth, as if she were plugging herself up, and walked to the trough. Anika stepped outside and came back with a bowl of water. "Here," she said, and Frida began cleaning her finger. It stung when it hit the water.

After she'd returned to the table where the mound of garlic awaited, her finger smarting but no longer bleeding, Burke leaned over and whispered, "It's the blood."

"What?"

"It bothers Anika."

"I don't understand."

Frida didn't get any further explanation because Anika was announcing something else now, about how someone had evidently neglected to soak the beans last night, which meant they didn't have enough protein for tonight's dinner.

Frida wished, suddenly, that Cal were here to witness what had just happened. She had barely thought of him all morning. Last night, before falling asleep, she'd entertained a flittering fear that without him with her during Morning Labor, she might totally lose it. They rarely separated, and when they did, it wasn't to go off with strangers. Neither tried anything new anymore. There was too much at risk.

The truth was, the morning had been wonderful. She could

be apart from Cal for a couple of hours. She could say what she wanted and be chummy with Fatima, without his disapproving gaze following her. She and Cal had separated for a few hours, and she had survived.

But now she wanted him with her. She felt purposeless without him. She tried to imagine what he was doing at this very moment. It was warm in the kitchen, but the sky outside was white and covered with gray clouds, and the trees beyond the Spikes were shuddering in the wind. It was probably cold, and Cal's hands probably hurt as they mended a fence or hammered a nail into a plank of wood. Maybe he imagined he'd been transported back to college, to those morning assignments. Did he feel comforted, doing that work with others? She knew those two years at Plank had stuck with him and that he held the memories deep inside himself. He coveted them, even, as if they were just beyond his reach.

Frida was glad when Anika told them they were almost finished. She wanted Cal.

Fifteen minutes later, she sat in their room, eating a bowl of mushy carrots, waiting for him. The group had invited her to eat with them in the dining room, but she'd declined. She knew Cal would rush to the bedroom when he was finished.

"You're here," he said as he entered. His shirt was dirty, his hair wet with sweat. He looked at her bowl of food. "Can I have some of that? I'm starving."

"What did they do to you?" she asked, moving onto the bed so she could sit behind him.

He'd been with a crew of about four others, dismantling the wall of bricks by the Bath. They needed to break it apart without damaging the bricks, and it was hard work.

"They're going to reuse them?" Frida asked.

Cal nodded and took a bite of food. The carrots were cold and bland, and he wrinkled his nose as he chewed. After he'd swallowed, he said, "They need a new outdoor oven." For the last few weeks there'd been a lot of debate about the oven, as it meant taking apart an original structure. "But I guess functionality trumps nostalgia." He held up his hands, their palms dyed reddish brown, his fingers chapped. "All I know is that job was a bitch."

She pouted and kissed the back of his neck.

"Aren't you being sweet," he said, turning around.

"Is that hard to believe?"

He raised an eyebrow and tried to hand the bowl back to her. She shook her head. "You eat it. I've been around food all day."

"Was it fun?" he asked.

"It was. They might let me bake bread."

"Really? That's great."

Did she hear a snag of mournfulness in his voice? Maybe he was thinking of the bread she used to bake him when they had first started dating, and the pizza bagels he'd beg for. "One of those and a blow job, pretty please," he'd said once, when she asked him what he wanted for his birthday. Or maybe it wasn't quite as precious as all that. Maybe Cal knew, as she did, that once she started making bread for the Land, she'd never want to leave.

He set the bowl on the floor and then sat facing her on the bed. He put a hand on either side of her skull, cupping it. Frida had once seen an old man do that to a pregnant woman at a bus stop. Frida didn't shake Cal off but let the weight of his palms rest there; maybe the brick dye would chalk off in her hair.

"You okay?" he asked.

"I cut my finger today."

She held out her index finger to show him, and he let her

head fall so that he could take her hand. He kissed the wound.

"Poor thing," he said, and kissed her finger again, and then her wrist.

"Can you believe we're here?" she said then, her stomach growing warm. Even now, Cal could make her core heat up like she was the center of the earth.

"I can't," he said, her finger in his fist.

She moved toward him, and the smell of his sweat hit her. That, and the unfamiliar dust of the Land. She liked the surprise of this new smell; she wanted it. She bit her lip.

Cal pushed her gently onto the bed and breathed into her neck, pushing his body against hers. In moments he had scooted her dress above her waist. The leggings she wore beneath belonged to Fatima, and Frida was afraid he'd say something, but he didn't, he just grabbed at the elastic waistband with an urgency she hadn't seen from him in a while. Maybe he liked the unfamiliarity. His eyes were closed; was he imagining someone else's body beneath his own?

She put a hand on his chest and said his name. He opened his eyes. She pushed him off of her. "Look at me," she said, and began pulling off his shirt. Despite the strange room, and the awful uncomfortable bed, and the secrets they'd kept from each other, Frida felt her desire for Cal expand and expand.

They didn't bother with foreplay much anymore, those courtship niceties of kissing and petting before they were totally naked. If Cal was going to kiss her deeply, or put one of her breasts in his mouth, Frida wanted him inside of her as he did so. They were married, they were efficient: they'd done this dance dozens of times before, they both knew the song.

As they moved together, it felt better than it ever had. This, she thought. She wanted to call out, but she bit her wrist instead, her whole body pushing. Cal had kept his eyes open, he was watching her, he was witnessing the pleasure she felt, and she knew he felt it, too.

"My . . ." she said, but she couldn't finish the sentence, whatever it was going to be, she didn't know.

Cal nodded. "My . . ." he whispered back.

He had lifted her hips toward him, and they were right on the edge of the glorious cliff when she closed her eyes, and her mind flashed to the moment the knife cut into the skin of her finger. Maybe it had felt good, the blade breaking the skin the way a boat parts water. Maybe it had been beautiful and clean like that. Cal was pulling her body around his own.

Suddenly she saw them yesterday. Micah was moving from the kitchen into the dining room, their first meal in there, and she couldn't stop looking at him. That long beard, and that raw patch of scalp on his head she didn't yet know about. Someone must have rubbed alcohol there first, right, before they sliced that piece of him away?

Her brother was sitting at the table before that bowl of soup and then the knife was cutting through her finger, the blade smooth and sharp, and Cal was now heavy atop her, groaning. He was saying her name, and she felt a pang of pleasure so bright it almost blinded her insides. She saw the coyote, it was standing there in the dining room as they ate quietly, its mouth dripping with viscera, and she shot her eyes open. *Look at Cal,* she told herself. The pleasure was receding like a tide. She had to bring it back. Cal kissed her, and she held him to her lips, as if he could suck out the images in her mind. But he couldn't.

"My . . ." Cal whispered as he came, but Frida said nothing.

Afterward, still naked, they lay on the bed, breathing hard. After a moment, Cal sat up and began sifting through the bed-sheets for his clothes.

"I need my stuff," he said. This morning he'd had enough of

being the Official Pussy Inspector and broke down and asked Sailor (not Micah, Frida noticed) for a shirt. Sailor had actually pulled his own T-shirt over his head and handed it to Cal. "We'll trade," he said. Cal had been wearing Sailor's slightly tight shirt ever since. It puckered at the armpits.

"Why don't you just let my brother give you some clothes? They'd fit you better."

"We have to go back home, Frida," he said, placing a hand on her hip.

"You mean to pick up more of our things?" She felt her body tense beneath his touch.

"For now," he replied, moving his hand. "But we can't just not make a decision."

"They're the ones voting," she said.

"But we have a choice, too."

She was silent.

"Today, in the kitchen," Cal said, "did you get an idea of where they're getting all their food from? I mean, did you get to look at their gardens? Where are they storing everything? Did they have any out-of-season fruit or"—his voice tipped—"anything canned?"

"I wasn't on a recon mission, Cal." She sat up. "God, could you please just let me have a few days to be here? With other people. With my brother?" She closed her eyes quickly; no doubt Cal had noticed that Micah had left Sailor to take care of them this morning.

"Don't," he said. "I don't want to argue."

He moved into a squatting position on the mattress. He was still naked, and it made her laugh. He almost tipped over, then righted himself, like a surfer.

"What are you doing?"

"Look out the window with me," he said.

They both perched on the scratchy mattress, hands against the wall and headboard for balance. Cal pulled the cheesecloth from the window, ripping it off its staples.

"Cal!" Frida whispered, but she couldn't help but laugh again.

They looked through the square of window. The scent of animal shit—or was that human shit?—wafted into the room. The air outside was cool, but, judging from how sweaty Cal had been after Labor, she knew it had to be warm in the sun by now. Frida leaned back and stuck her arm through the window. She put her palm against the side of the building, which was hot to the touch and rough as a pier and gritty with dirt.

Their room was on the north side of the Hotel, and from the window they could see beyond the main street to the areas Sailor and others had alluded to since their arrival, but which Frida hadn't yet been curious about. Until now. The space was wide open as a meadow. It was mostly free of trees, except at the edge, where things grew wild and uninviting; a Spike rose menacingly above this patch of untended land and, next to it, another lookout Tower. Someone must be on duty, Frida thought. She wondered if they ever trained their binoculars in the other direction, toward the Land's inhabitants.

To the left was the showering and laundry area, where clothing hung like prayer flags on multiple lines stretched between four trees. Frida watched as a man walked naked from one of the shower stalls to the lines. He grabbed a pair of pants hanging there and put them on.

Across the field was a structure that looked as if it had been recently constructed, perhaps out of materials collaged from various ghost-town buildings and whatever else the Land could get its hands on: the wood was both old and new looking and placed side by side; the planks gave the building stripes. The roof was made of corrugated metal and held secure with tires and wire, like their shed had been. The doors were tall and wide, like a barn's, and a man came walking out with a goat on a rope. Along the outside of the building were animal pens.

"Is that where August's mare lives?" Frida asked.

"I assume."

"Where is he, you think?"

Cal shrugged. "You should ask Micah. He'd tell you."

"I doubt it."

"See if you can get him alone."

"I want him to come to me. He's been so cavalier about seeing me, after all this time. He just left us with Sailor this morning." Frida felt the tears coming, and she tried to laugh them away. "Jeez, I guess the hormones have arrived."

Cal leaned into her. "You deserve to spend time with him, Frida. He's your brother. Just ask him."

"I'll try."

It felt good, Frida thought, to be talking like this. They were plotting again; they were on the same side. They had returned to each other. They were something the world could understand. This had been how she'd imagined it, when Cal had first asked her to leave L.A.

———

Over the next two days, Frida began to get a handle on things. The lingo, for one. Residents on the Land didn't work; they *labored*. They didn't garden; they *farmed*. And those Spikes that surrounded them? They were called *Forms*.

Learning these terms gave Frida a thrill. It was easy, like learning pig Latin or the gibberish she used with her friends as a girl.

Her new Land friends couldn't keep her out of conversations for very long—not that they were doing it on purpose; it was just the way they spoke about their world. The vocabulary was so simple, it was impossible not to start using it.

Frida was officially out of her half coma. After all, the Vote wasn't far off. Not that she and Cal had talked about that. He was too busy asking Micah question after question to notice that his wife was campaigning.

If anyone noticed what Frida was up to, it would be her brother. Nothing ever got past him, never had. He'd always seemed to see her for what she was.

Was that still true? There was something weird about Micah now, and not only that he was alive when she'd been grieving him for the past five years. He didn't participate in Morning Labor, for instance, nor did he seem to have a security shift, as far as Frida could tell. At dawn, when everyone else was headed to work, he disappeared with a handful of others, all of them men, Frida noted, including Peter. (*His cabal of yes-men,* she imagined Cal saying, but she didn't dare bring this up with him.)

No, Micah was odd because he could send people away with a distracted wave without seeming like an asshole and because he hadn't yet asked after Hilda and Dada, not really. He hadn't yet asked to be alone with her, didn't seem interested. As if it hadn't occurred to him.

Then, on their fifth day on the Land, Micah came into the kitchen at the end of Morning Labor. "Want to go on a walk?" he asked her. Just like that. A few minutes earlier, her fellow cooks had told her that housekeeping had been sent to clean out Sue's stable, which meant August would be returning soon. He was a few days behind schedule, though no one had any idea why. Or they did, but they weren't telling Frida.

When Micah walked into the kitchen, the group got giddy. Not Fatima, who, Frida knew, spent a lot of time with Micah. But the others, even Anika, seemed to speed up their movements and speak louder, come into focus, into high definition.

"Can Frida leave early?" Micah asked Anika, who nodded and took Frida's knife out of her hand. Strands of purple cabbage hung like party streamers from the blade.

"Leave it to me," Anika said. "I'll clean it up." This was not the same woman who had gotten all huffy about the beans on Frida's first shift.

They left through the back door. Construction on the new outdoor oven would begin soon, perhaps beautifying this dry lot of soil behind the Hotel. Until then, there were only the outhouse and a large fire pit dug into the dirt. The day before, Anika and Burke had used the pit to roast rabbit; the animals had been caught in rusted-out traps that almost everyone was worried about. Burke claimed that the contraptions wouldn't last through the year, and Fatima had accused him of being an alarmist.

Morning Labor ended at lunch, but Frida had recently learned that some people spent the afternoons rotating through optional jobs like hunting, foraging, trapping, and composting. There was also construction of the Forms. Frida had wanted to join that group, but Sailor said it was by invitation only. He told her they could help with plumbing, if she wanted, which meant getting rid of human waste: cleaning bedpans, digging new latrines. "Everyone's favorite," he said. If not, she should just take it easy. He said most people had their afternoons off. "We value leisure time here," he said, "and the boredom of a slow life."

Weeds scraped at Frida's ankles as she followed Micah. It was still warm out, and Fatima had lent her a pair of sandals. They were a size too big, though, and with each step they slipped off her ankles and slapped the ground, bringing up dust that settled under her heels. Micah seemed to sense she was having trouble keeping up and slowed down.

"I was afraid we'd never get to hang out," she said when she'd reached him.

"Really?" He turned to her. "I didn't mean for you to think that." He pointed across the meadow to the untended land that bordered the Spikes. The Forms, she reminded herself.

"Let's head for the shade," he said.

"Are you going to murder me there?"

"Looks spooky, I know," he said, with a laugh. "But I promise, they're just like any other woods around these parts."

They passed the large garden, which had all the same vegetables she and Cal grew, though far more of them. There were also kale and a kind of squash she didn't recognize and what looked like a persimmon tree, lying dormant. She saw Sailor bent over the plants and wondered why Cal hadn't volunteered for that team.

"I hear your husband's on the brick headache," Micah said, and tapped his forehead, making his straw hat wobble. It looked like his beard had been trimmed, if only by an inch or so. Or maybe it was the same length, and for once it had just failed to shock her.

"I think they're almost ready to build the oven," she replied.

They kept walking. A group of people was heading with buckets to one of the Land's three wells, and Micah waved to them. She guessed they were on housekeeping.

Once they'd passed the camping area, where people could come to sleep outdoors, Micah asked, "How is it?"

"How's what?"

"Being married?"

"It's good," she said. "I mean, I don't exactly notice it any-more."

"It can't be as natural as breathing."

She wasn't sure what he was getting at. "No, I guess not. But we've been together for a long time. You know that."

"It was just you two out there."

Frida nodded. Should she bring up the Millers?

They had reached the lip of the woods. There were Spikes—Forms—bookending this section of trees, and probably beyond it, too far away for Frida to see from where she stood. Micah stepped aside and swept his arm across the space before them like a circus ringmaster.

"Cal's dream come true," he said.

"These woods?" Frida said, but she already knew what he meant.

"No. The two of you, the end of the world."

"What does that mean?"

He shrugged. "Just that Cal has always preferred you above all others."

"Doesn't everyone?" She walked ahead, into the trees, as if she weren't afraid of the darkness there, the branches that might cut her.

He only let her lead the way for a few seconds, and then he moved in front of her. As with Sailor in the Spikes—*The Forms,* she told herself again—Micah somehow knew his way through this dense forest. If you looked hard enough, if you were will-ing to step over bushes and dead trees, you could discern a path.

"Is August coming from this direction?" Frida asked.

He shook his head. "He can't get the carriage through, so he has to go around."

Micah was wearing the same green Polo shirt he'd had on the day they'd arrived, and Frida tried not to look too hard at it. It would only bring back that first day again, and she couldn't revisit that. If she did, she might lose her footing or stop speaking or hyperventilate. Sometimes breathing wasn't natural. Instead, she kept her eyes on a piece of loose straw, flapping from the brim of his hat. Perhaps because she couldn't see Micah's face, she felt emboldened to ask him questions.

"So August told you I was out here," she said.

"He returned from the last trip and said, 'You won't believe it, Mikey.'"

"Did you? Did you believe it?"

Micah didn't reply, and she couldn't even guess what his reaction had been. Before, she'd been hurt that he hadn't come to see her, but now she felt angry. She deserved answers.

"This way," he said, and pointed up.

She thought for a moment that he was asking her to shinny like an animal up a tree trunk and was about to tell him she didn't have the upper-body strength for such shenanigans, when she saw pieces of wood had been nailed into the trunk. A little ladder. Someone had built a wooden platform in the tree.

Micah made a basket with his hands and knelt. "I'll give you a boost."

"Is this the clubhouse?"

He stood up and sighed. "No, Frida, this isn't the clubhouse. It's just where I go to clear my head." He held out his hands again. "I just thought you might like to see it."

The wood steps were smoother than she expected, as if some Land member had buffed them before nailing them into the trunk. If Micah had been the kind of little brother who liked sports and played war and broke bones and heads off Barbie dolls, this might have felt like a return to their youth. But as a kid, Micah had preferred to be alone, preferably indoors. Sometimes he could fall into a stormy mood, but if you left him be, he'd cheer up eventually. He liked to read books and take apart the toaster and post videos on the Internet about their bathtub. At eight he read about the sixteenth-century seaman Martin Frobisher, who discovered Canada and later fought off the Spanish Armada. Micah became obsessed with him and for months asked everyone to call him by that name. Nobody did, not even Frida, who usually put up with him. Hilda just laughed it off and said, "My children: the greatest mystery of all."

While Micah was being a nerd, Frida would roam the neighborhood, hiding in people's backyards, pretending she'd run away. Once she'd broken into the yard directly behind theirs, just for kicks, and had accidentally stepped on a tortoise. The house was a freakin' menagerie, Dada said later. The animal's shell was warm against her bare feet, and solid, but there was the knowledge of a soft body beneath it, and Frida had screamed. Micah happened to be in their own backyard at the time, and when he heard her, he stuck his head over the fence and said, in the beleaguered voice of Hilda, "Come now." He had just turned six.

Micah's grown-up tree house was an open wooden platform with the trunk in the middle and a single railing around the edge, as if fighting off the branches. Frida had never been on a boat, but she felt like this was what it must be like, standing at the helm, the water beckoning and teasing and scaring below. She didn't think she was afraid of heights, and this tree wasn't very tall, but it had been a long time since she'd been above anything, even a canopy of leaves, and she held on to the railing with both hands.

They sat on two collapsible camping chairs, and from a plastic toolbox, Micah pulled out a cloudy glass bottle and two creased Dixie cups. The cups had begun to collapse in on themselves, and Frida could tell they'd gone from soggy to stiff and back again multiple times.

She nodded at the bottle. "What's that?"

"I'd call it whiskey, but then you'd be disappointed."

"You guys make liquor here?"

He shook his head. "We traded for it."

"Who makes it?"

He raised an eyebrow and poured the alcohol into the cups, which sagged with the weight of the liquid. "Please don't give me the Cal treatment, Frida. All day, people are asking me questions, wanting my advice, asking for solutions. And then, on top of all that, your husband comes along with an endless questionnaire. I just want to relax."

"Oh, fuck off, Micah. I just spent the last five years thinking you were dead. And here you are, playing king in a tree house. You don't get to relax."

Her brother looked skyward, as if a response might be written in the tree branches above. "You make a valid point," he said, and handed her one of the cups.

She took the cup, but she didn't drink. Just one sip wouldn't hurt the baby, would it?

"A bouquet of lighter fluid and piss," Micah said, and downed his.

She put the cup to her nose. Being drunk actually sounded wonderful, and the sharpness of the liquor was as pleasing as it was revolting. The burn traveled through her nostrils and into her throat.

But she couldn't.

"I'd rather not," she said.

"Seriously?" he asked. "You? Turn down a drink?" But he was already putting down his empty cup and taking hers. Between his fingers, the edge of cup folded into a triangle, threatening to spill its contents. Micah brought it quickly to his lips.

"I have so many questions," she said.

"Ask them, then," he said. "But up here, there's no need to be a mouthpiece."

"'A mouthpiece'? You mean Cal's?" She leaned back. "Don't be typical, Micah."

The phrase was out before she could even think about it. Hilda used to say it to him when he'd refused to eat dinner with them or put on shoes to go to the market. Or when he'd say something witty and cruel, his mouth curved mean and smug.

"I want to know just as much as my husband does," Frida continued. "Why wouldn't I? It's only natural."

"I can't believe you just said 'Don't be typical' to me."

"Now you know how it feels to see a ghost." She smiled. "Brings you back, doesn't it?"

He looked bashful for the first time since she'd arrived.

She picked up his empty cup and swiped at its bottom with her index finger, then brought her finger to her mouth. The liquor tasted sour as vomit.

"Ugh," she said. "You must be really desperate to drink this."

"Maybe," he replied. "You know how it is, not having all the things we used to have."

"Tell me about it!"

Cal never wanted to gripe about what they were missing from the old life. He said talking only made the loss more palpable, the absence more glaring. He said it was a form of self-punishment.

Micah seemed to agree with Cal, because he didn't go on. Instead, he took off his hat and ran his hands through his scraggly hair. Frida was glad she couldn't see the top of his head, that bare spot. Micah looked more like her with his hair long, and she realized she was proud of this. It would help with the Vote.

"I want to know why you did it," Frida said.

Micah raised an eyebrow.

"Stop with that phony face," she said. "Tell me."

He sighed. "Look at this way: no one's looking for me, are they? The police, Homeland Security, they were idiots. Or, I don't know, maybe they were just underfunded. They got a piece of me, tested my blood. They had a piece of clothing my poor family could identify. They had enough to close the case, and they did."

"'My poor family.' Listen to you. There have to be easier ways to disappear."

"I was the head of the Group by then. One of them, anyway. You had to have known."

Did she? She supposed she should have. "So what?" she said.

"What I did, or what I pretended to do, proved we were serious. Not only to you and everyone outside, but to our own members, the little shits who'd started skulking around only because they'd heard we might feed them." He shook his head and put his hat back on. "My stunt proved we were in control. For the first time, people were scared of us, really scared of us. Until that day, no one important cared about what happened outside the Communities."

"I hate to say it," Frida said, "but they still don't care. But maybe you knew that all along, and that's why you didn't actually commit suicide. Maybe you were too chickenshit to do it for real."

He smiled. "That's beside the point, don't you see? My stunt, whether real or not, freaked out the Communities, and it got us new members. Good ones. People saw we could be powerful."

"That's one word for it."

"Within a month we'd expanded our encampment by a mile."

"But what about the guy who really did blow himself up? He died anonymously for your . . . cause? Just like that?"

"Better that than to die pathetically, ignobly." He looked at her. "Isn't that how Hilda put it? I read the websites. I read what she thought of me and what I'd done."

Frida felt the old anger feathering in her chest. "You know she came to terms with it. They both did. They had to."

"I suppose," Micah said.

They were silent. Micah's words filled Frida's head—*my poor family.* That was all they were to him. Three people he could dupe.

"Frida?"

"How are they? Do you hear from them?"

"Ah," he said, grinning. "See? That's what you really want to know."

"Just tell me. Are they okay?"

The day Hilda and Dada moved to the Group's encampment, Frida said she was disgusted. "I know you're scared out here, I get that," she'd told them. "I know they've promised to keep you safe, that you won't have to worry about money. They'll probably treat you like royalty. But they took Micah from us. Doesn't that matter?" When her parents wouldn't answer, she told them of her and Cal's plan. "We're getting out of L.A. as soon as we can."

"Don't be stupid," her mother had said. "You have to stay."

That's when her father had called her a traitor, for leaving willingly. Frida didn't say that Micah was the real traitor. She wouldn't.

Her mother had hugged her goodbye and said, "Enjoy the air out there." Her father had hung back, saying nothing.

The Group had welcomed Hilda and Dada to join them and partake in their resources. They would make sure they were safe and that they'd never go hungry. After all, their son had died for the cause. All the Group asked for in return was the house, which they'd dismantle for parts, and the land, which they'd use for who knew what. Frida didn't want the property; she didn't care about inheritance and all that. The pain she felt at their leaving for the encampment was about something else. She was losing everyone. Cal had been trying to convince her to leave L.A. for months, but it wasn't until her parents told her of their plan that she agreed to go. There was no reason to stay.

"I haven't been in touch, if that's what you're wondering," Micah said. "But the encampment stretches to downtown now. And there's another one planned. This time, near the beach." He paused. "I'm sure they're fine. Better than fine."

"How can you say that? They think their son is dead."

Micah said nothing, only stacked the cups and put the cap back on the bottle of liquor. There were only a few drops left; it would be empty in one tiny sip.

"You could have gone, too," he said. "It took Hilda and Dada two years to agree to it. I would've thought you'd join them right afterward."

God, she wanted to shove him off the tree. "I would never."

He raised an eyebrow. "You never did live in reality, Frida. Or maybe I'm wrong, and that's more Cal's problem."

"Leave him out of this."

"Frida."

"What?" She hated him saying her name.

He was looking right at her.

"What I did, my disappearing, it wasn't selfish."

"Sure, it wasn't. You had a cause, you said that already."

"No," he said. "Well, yes, but also Hilda and Dada are comfortable now."

"I don't want to talk about this anymore." She couldn't

describe to him how it felt to have first her brother taken, and then her parents.

"What else can I tell you?" Micah said after a moment.

"Can you guys procreate?" she asked. "Are the women infertile?"

"That's what you want to know?" He laughed. "Wow, Frida. I never knew you were such a geek. You like zombie movies, too?"

"It's a good question, and you know it." This was the moment to tell him she was pregnant. *Do it,* she thought. But she couldn't even open her mouth.

"We can procreate, yes. But that doesn't mean we do."

"Let me guess," she said. "You believe in containment."

"Don't make fun of our brand."

This time, they both laughed.

"The containment stuff . . . does it have anything to do with blood?" she asked. "Like, you know . . . rejecting it?"

"Why?" But then he held up his hand to keep her from answering. "It's not blood that's the problem." He paused. "It's the color."

"Red?"

Frida remembered Sandy. That first time, meeting by the creek. How Sandy had snatched Cal's red bandanna from Jane. And, later, how Sandy had turned away from Frida in the shed so as not to see the red sleeping bag.

Like Sandy, Anika was afraid of a color. How had Frida not put that together?

Frida looked at Micah. "Why is she afraid of it?"

Micah smiled. "It's a thing. She has negative associations."

"What does she associate it with?"

"Pirates," Micah said, and Frida reared back.

"What?" she said. "What do you mean? They're real?"

"They're nothing to worry about," he said. "They were only a problem for the original settlers."

So Sandy had to be an original settler. Bo, too.

Frida held her voice steady. "Tell me about Bo and Sandy."

"I don't think that's a good idea."

"So you don't deny knowing them."

"I don't want to scare you away." He smiled. "Not yet, at least."

"That's all you're going to tell me?"

"For now."

Her brother stood and held out his hand to help her up. His hand, she noticed, was unscathed, uncalloused, unworked.

"You're the boss here," she said.

"Somebody's gotta be," he said, and shrugged.

Once she and Micah were halfway across the field, Frida said, "I have a present for you." She wanted him to wait while she ran up to her room. "I want you to unwrap it outdoors."

"You mean the baster?" he said.

"You already know about it?"

"Nothing gets by me, Frida. That much should be clear by now."

The look on his face. Years ago, when he announced to their family that he'd applied to Plank, he'd had a similar expression. There was a deliberateness to the look, a purposeful arrangement of his features, an anagram of emotions. If Frida stared at him hard enough, might something entirely different be revealed? She thought she had uncovered the old Micah when they were in the tree, talking freely, but she'd been wrong. He had himself under control. Frida couldn't get to him.

12

Cal hadn't taken a shower this good in years. He and Frida had never been able to get this much warm water on their own, and he'd never considered how comforting even a rudimentary wooden stall could be. He could've been in Cleveland again, showering in their cold moldy bathroom while his mother cooked breakfast in the kitchen. She'd be frying up the eggs his father had dropped off the night before. Cal leaned his head back, and the water fell across his face.

The reverie didn't last. He couldn't stop thinking about the fact that Frida had talked to Micah the day before. She'd told Cal very little about it, just that it felt weird, hearing Micah talk so openly about the man who had died instead of him. "What happened to my brother?" she asked as they fell asleep, and then, "Why is he like that?" They weren't questions anyone had answers to.

He thought it would make him happy that Frida was finally seeing the truth about Micah, but he was surprised by how much it unsettled him. Her optimism was fraying. She had always believed people, especially her family, were good, that the world would only allow so much suffering. In the past, some of that delusion (because wasn't it delusional, to carry on with such thoughts, after all they'd seen?) must have rubbed off on him. He hadn't realized how much more palatable she'd made their days.

If she suspected something of Micah, Cal could barely stomach the thought of him.

Cal wanted to know what Micah thought about her pregnancy. Did the prospect of new life, of a new family member, soften him? Probably not. Did it do just the opposite? Cal waited for Frida to tell him, but to these questions all she'd said was "August will be coming back soon." As if this were news, as if she'd done useful detective work. She wouldn't be giving Micah the baster, she said. "He doesn't need it," she said. When Cal asked her when she'd told August about her brother, she said she couldn't remember. "I guess it just slipped." So much for their agreement to keep the past a secret.

From the shower, Cal heard someone squawking like a rooster at dawn, and then the crunch of dirt traversed by wheels. If he didn't know any better, he would have imagined a truck passing just out of sight, imagined the weight of its body and the heave of its motor as it pulled up to the barn. Because it was just lame nostalgia, he would never admit it to Frida, or to anyone, really, but sometimes he missed the sounds of large, gas-guzzling engines: idling and accelerating, their gruffness and soot. Childhood sounds.

He didn't go to investigate the sound because the water from the old plastic jug was almost out, and he wanted all of it. It felt great. They'd been on the Land for almost a week now, and he deserved this shower: Morning Labor had been kicking his ass. They had finally started on the outdoor oven. He and the others had carried the bricks to the lot behind the Hotel and then dug out the area where the oven would be built. His neck and arms were sunburned, and his hands were chapped as badly as they'd been when he and Frida had first found the shed, when there'd been so much to build and do outdoors. At least back then, she'd kiss his hands every night before bed, blow her cool breath on his open cuts. Now she didn't offer, and it felt pathetic to ask.

Morning Labor wasn't as trying as the discussion of it; there was a strict protocol to follow with any new project, and the members on his team were nervous about taking a wrong step. He didn't want to use that word, *team,* but everyone else did, and it had seeped into his vocabulary when he wasn't looking. A woman named Sheryl had forced them to measure and remeasure the spot planned for the oven, to ensure it was the one decreed. *Decreed* was Cal's word—his team had assured him it'd been a group decision, but he didn't believe them. Cal had seen Micah and Peter talking in front of the Church. It was a meeting, Cal realized, by the way their voices dropped low, their faces no longer playful. They were the ones making the decisions.

Cal could ask August about it himself. That sound must be him arriving, wasn't it? Cal realized it as soon as the water trickled to a drop, and another drop, then nothing. He hurried out of the stall and shook himself dry before throwing on his pants and Sailor's T-shirt. Cal had been told he could grab anything from the line that fit, but he refused. He knew he was being petulant—even Fatima had used that word to describe him to his face, smiling as she did so—but he couldn't help himself. He didn't want to leave his own pants for a stranger. The longer he stayed on the Land, the more possessive he became.

By the time he reached the barn, his still-wet skin had stained his clothing dark, and his pants were making his legs itch. He should have used one of the drying rags, old tablecloths and bolts of linen that the Land used as towels, but he'd refused that as well. He *was* petulant, wasn't he? He was stubborn as a two-year-old. If Frida saw him, she'd laugh, ask him if he'd peed his pants. But she'd gone foraging and wouldn't return for a while.

She was out making friends, volunteering for extra work. Tomorrow morning, an hour before Labor began, she was meeting Anika, her team leader, to discuss bread making. If she was suspicious about Micah, she didn't seem to carry those feelings to the people she cooked with, at least not outwardly. Cal tried to be happy about this; his wife hadn't become someone else entirely.

August's mare, Sue, had already been led into the barn, but otherwise, everything else was as Cal expected it. August looked as he always did, standing there next to his buggy: same gray sweat-suit and combat boots, same wraparound sunglasses, same beanie covering his head. Cal held up his hand as a greeting, and August simply nodded, as if this were an everyday occurrence. This, too, Cal had expected: August's capacity to remain unfazed, no matter what.

People were gathered around him like eager children, and Sailor had climbed onto the edge of the buggy, leaning in to get a better look at what had been collected. This did surprise Cal: someone else besides August was allowed to touch the cart.

Micah stood off to the side, and Cal saw that he was watching him. Had he taken note of Cal's brief moment of shock? Cal hoped not. He pulled at his wet T-shirt, fiddled with the scratchy waist of his pants, and kept walking.

"So he's back," he said as he reached Micah. "It's quite a welcome."

Micah nodded. "Always is. August comes bearing news and gifts. And Sue's our mascot, if not one or two men's soul mate."

Micah held up a hand, gesturing for August to join them. "Plus, he's got your stuff."

"Ha," Cal said, but as he did, Sailor lifted a large duffel bag out of the cart. It was the purple bag with the teal straps, the one Cal and Frida kept on the highest kitchen shelf. It was now stuffed as full as it had been when they'd left L.A., long and heavy as a dead body, a mafioso joke too obvious to make.

August yelled at Sailor to put the bag down, and Sailor complied immediately. Dave pulled him off the buggy, yelling, "Come on, you nosy motherfucker!" They were laughing.

Pulling off his hat, August jogged over to Cal and Micah. His head was bald and shiny with sweat, but Cal thought he could make out a vague shadow of hair growth—a receding hairline. August would probably go to the Bath soon, take care of that right quick. Someone would probably volunteer to shave it for him.

"Cal," August said, and shook his hand.

"You broke into my house."

"This guy," Micah said, looking at Cal, "has no time for niceties." He put a hand on Cal's shoulder and gave it a friendly shake.

"Let's go talk," Micah said. "I'll get Peter."

"Sounds like a fine idea," August said, and put his cap back on.

"Which one of you okayed the theft of my property?" Cal asked.

"Your *property?*" Micah said.

August shook his head and pulled off his sunglasses. Cal sucked in his breath.

But they were just eyes. Dark brown eyes. August looked less intimidating without the sunglasses. He must have known it, and that was why he had removed them.

"Come on, Cal," he said, blinking in the sunlight. "Give us a break. You gotta know, we're not out to get you."

"You need clothes, don't you?" Micah asked.

"I had an extra shirt and a pair of jeans when I arrived," Cal said. "Sailor returned my flashlight and sleeping bag, but he didn't know what happened to the clothes. Said I should holler

202

if I see someone wearing my stuff." It was almost too absurd to make Cal angry anymore.

August took in Cal's too-small, soaked shirt. "Cal. You're a man." He paused. "Sailor, he's . . . I don't know. A boy? A kid. You can't be wearing that, it doesn't fit."

August started to laugh, and so did Micah. Cal waited.

"Sailor, get the bag!" Micah yelled, once he'd caught his breath. To August and Cal he said, "Follow me."

They did as he asked.

They walked to the Church. On the way, August asked Sailor to go find Peter as soon as he'd dropped off the bag. Sailor nodded urgently and said the bag wasn't that heavy, that he could do both in one trip.

"He loves to take orders," Micah said once Sailor had taken off, duffel slapping at his side. "He's the only one who actually likes Morning Labor."

Cal had been about to make his own snide remark. He wanted to ask Micah why he didn't participate in Morning Labor, but he didn't because Micah would most likely reply with something cutesy, something like, *We work, too, just with our heads.* A sad disregard for manual labor, though that would be strange, considering what they'd learned at Plank: the field *and* the book, a symbiotic relationship. Perhaps Micah, in his grab for power, had disregarded half the skills that had led him here.

Cal told himself he wouldn't give his brother-in-law the satisfaction of clever answers. He would withhold all questions. Perhaps if he seemed uninterested, they'd be more willing to explain how everything worked.

Wasn't that, in the end, what he wanted? To discover how this place worked—not just its outward system of organization but its inward, private one as well? Its secret machinations, the strings that gestured the puppet. Who was the puppet, though? Maybe it wasn't all that sinister. Frida was probably right; he was descending into paranoia. Maybe it was more like a car: just lift the hood, and you'll see how everything works.

The Church was cool inside, the empty pews gathering dust in the sunlight. The studio lights, tall and spindly as prehistoric insects, waited nearby, disturbing but, for the time being, powerless.

Cal wanted to go to the second floor. He didn't realize this until Micah hoisted himself to the edge of the stage, and August slid into the first row of pews. There was no way this was where they conducted their meetings each morning. There was a war room upstairs; there had to be.

A few moments later, Peter and Sailor walked in. Peter was holding the bag now, and when he caught Cal's eyes, he lifted his chin, beckoning him to come retrieve his possessions. Instead, Cal sat down next to August in the pew.

"I'm working on the goddamned traps," Peter said. "I was about to tell poor Sail to fuck off, when he said Gus was back. He was dragging *this* along the ground." Peter hefted the bag onto the second-row pew.

"I think I hurt my back," Sailor said.

"Too much pussy inspecting," Micah replied, and August laughed.

"I wish," Sailor said.

Cal laughed, but no one else did.

"I guess I'll see you guys later," Sailor said then.

"You're not allowed to stay?" Cal said. Damn, a question. He couldn't help himself.

"Sure he is," Micah replied. "Have a seat, Popeye."

Sailor hesitated, but when August and Peter said nothing, he sat in the pew behind Cal.

"So, California," Micah began. He was swinging his legs, hitting the side of the stage with the backs of his heels. The wood was scuffed there; maybe this *was* where they conducted their morning meetings.

"So."

Micah stopped swinging his legs, as if this were a habit he were trying to break himself of. "August only let himself in."

"Theft," August said. "He used the word *theft*. He thinks I stole from him."

"Ah yes, he only *stole* your property because we knew you'd need more stuff. August returned from his original route only hours after you arrived. I told him to turn around and get things he thought you might want. Otherwise, I'm sure you'd convince Frida to go back home with you, if only temporarily."

"Why would that be a problem?"

"Because this isn't a place you can just visit," Sailor said.

"Sailor . . . " Peter said.

"It isn't?" Cal said. "Frida and I are stuck here?"

"Of course not," Micah said. "But it's dangerous to have you coming back and forth. Not many know I'm alive, and it has to stay that way. We can't attract attention with people waltzing in and out as they please and giving away our location. If you want to leave, it would be for good."

"I see." Cal imagined telling this to Frida; she would not take it well. "But August is always traveling the route, isn't he?"

"August isn't you," Peter said.

"What he means," Micah said, "is that August is the best candidate to trade with the few settlers nearby and to perform a regular security sweep."

"I don't know about 'the best,'" August said, "but when I tell people I'm a loner, they believe me. Or they assume it right off. I get special treatment." He brought an index finger to his cheek and tapped twice.

"Wait—why?" Cal said. "Because you're black? That's ridiculous."

"What the fuck are you talking about?"

"I only meant—"

August winked. "I'm just messing with you. Come on, Calvin—that's your full name, isn't it? I know you thought I was some kind of recovered addict. I put you on edge."

Peter, who was sitting in the pew across from Cal and August's, shifted his body so that his legs blocked the aisle. "We do things for a reason."

"I'm the last black man on earth—or at least around here."

"They should make an action movie about you," Cal replied.

This time, everyone laughed.

"There's more to it than that," Peter said. "People treat August special not just because of how he looks but because of who he is. He's very talented at getting people to open up to him."

"Amen to that," Micah said.

August shook his head, but Cal could tell he was pleased.

"I still don't like that you were in my house."

"Yours? That house used to belong to Sandy and Bo," August said.

"They're dead. I'm not."

Micah sighed. "We had to make sure no one else had been there since you and Frida left. And, besides, we were curious."

"You never invited me inside," August added.

That was true. Why hadn't they?

Micah pushed himself off the stage. "Cal, we're happy you're here. I mean, it's crazy."

"It's unprecedented," Sailor said.

"No other outsiders have been allowed in for a long time, you understand?" Peter said. "Sailor's right, it *is* unprecedented."

206

Sailor smiled.

"Do you think the others will want us to stay?" Cal said.

"I've taken it upon myself to, you know, ask around." Micah paused. "Everyone's supportive, but I'm considering delaying the Vote until everyone has had a chance to get to know you. I want everyone certain."

"What if we don't want to stay?"

Micah raised an eyebrow. "What if *you* don't want to, you mean?"

"You can't expect me to just accept this place blindly."

"I know you have a lot of questions," Micah said. "We'll answer them in due time."

No one spoke for a moment.

"We'd like you to come to our meetings," Peter said.

August and Sailor were silent.

"Why?"

"Because you're really smart," Micah replied. "Plus"—his voice grew soft—"you're my brother-in-law."

Cal looked back at his duffel bag. The handles were scrunched narrow where August, and then Sailor, and then Peter, had carried it.

"I don't know."

"There's time," Peter said.

"Nothing but time out here," August said.

"What about Frida?" Cal asked.

"What about her?"

"This is a boys' club, I gather."

"I suppose that's one way to look at it. We're the ones charged with keeping the Land safe—we're the most physically capable. And mentally, too." He paused. "If you want in, you need to follow our rules, keep our discussions private."

Cal nodded. "Sure, okay."

"He's serious, Cal," Peter said. "Don't tell anyone what we talk about."

This speech, Cal realized, was directed at Sailor as well.

"No spousal privilege," Micah said.

"You don't want Frida to know what we discuss?"

Peter shook his head. "Only who's present. The others have learned not to ask."

"It's no big deal, Cal," August said. "Most of it will be summarized during the Big Meeting later on."

"*Most* of it'?" He couldn't help himself.

"Cal," Micah said. He stepped forward and put out his hand.

August and Sailor were watching them, waiting.

Cal hesitated.

"What is it, California?" Micah had put his hand down.

"Things will really change," Cal began, "and in just a few months."

"They're changing already," Micah said. He crossed his arms.

Cal nodded. "It won't just be the two of us anymore."

"Yeah, there will actually be others around," Micah said. "Imagine that." His eyes were hard, but there was also a blankness there. He didn't get it.

"Didn't August tell you?" Cal said.

Everyone looked at August, who had put his sunglasses back on. "Tell them what?"

"That Frida's pregnant."

No one said anything. Someone walked by outside, dragging what sounded like a shovel along the dirt. *Scrape . . . scrape.*

"Is that right?" Peter said. His eyes were on Micah.

Scrape . . . scrape.

Micah didn't say anything. Sailor, for once, wasn't talking.

Cal waited, brushing his hand along the seat of the pew. His fingers came up dusty.

He had let himself be so stupid. Micah had said he was smart, but he wasn't. He was an idiot. He'd assumed that Frida had told August she was pregnant, and that August had told Micah, and that when Micah invited them to stay on the Land, he was inviting three people, not two.

But, of course, Frida hadn't mentioned the pregnancy to August. She hadn't told him about her baby, but about her baby brother.

"In her defense," August explained, "she was high as a kite when she told me about Mikey and his suicide." He smiled. "I gave her a Vicodin."

"Ah, Frida," Micah said, and snorted. "She always loved getting fucked up."

Cal was hardly paying attention. Frida had lied to him, and now these men knew it. He looked like an ass, keeping a secret without even knowing he was doing so.

"How far along is she?" Peter asked him.

Cal admitted he didn't know. "Not very far." He paused. "But she's happy about it, and so am I."

Again, no one said anything.

Cal stood up, the words *get out, get out* ringing in his mind. He was so upset with Frida, with her betrayal, he needed to be alone. "I have to go," he mumbled.

Without looking at Sailor, Peter said, "Help him with his bag."

Cal stepped across the aisle to retrieve it. Sailor grabbed one of the handles, and they both carried it out of the Church.

The bag was heavy. What possessions did August think they needed? Cal and Sailor carried the duffel toward the Hotel, where Cal expected they'd haul it onto the bed and unzip it—to find what, exactly? Cinder blocks and sneakers, maybe; a heartless joke, and here he'd been, despite his protests, lusting briefly for his gray sweatshirt and his khaki shorts that Frida had mended beautifully a few months back, so that they felt almost new. There was no way August knew what things they'd been longing for.

It was a short walk, but the main street felt endless when your arms hurt; Cal had learned this recently, carrying all those damn bricks. Apparently, the two wheelbarrows were in use elsewhere on the Land. "It's meditative to carry the bricks with our hands," someone from his team had claimed. From then on, no one had complained.

It felt like such a Plank thing, to take one's sweet time constructing something new and to value the hours ticking by. Cal had been annoyed; it fetishized the inefficient.

When he and Sailor finally got to the room, they dropped the bag on the floor. Sailor lingered in the doorway.

"I doubt you need my help unpacking."

"I guess not. There's no place to put things, anyway."

Sailor didn't move.

"Congratulations, man."

"Oh, thanks." Cal paused. "Should you be saying that?"

"There's no protocol for this kind of thing."

Cal waited. Sailor wanted to talk; Cal knew it. He wanted to introduce this world to an outsider, and if Cal waited long enough, Sailor might tell him everything.

"No one's had a baby here for a long time," Sailor said. "I've never seen it happen."

"And that doesn't seem weird to you?"

He shrugged. "This life—it's my second education."

"What was your first?"

Sailor held up his fist and knocked at the air.

Cal's voice caught in his throat. "You're a Planker."

"Last class. Well, would've been. Everything shut down after my first semester."

Shut down. Cal saw the farmhouse, and the fields, and the stove in the room of some lucky second-year, gone cold.

"How'd you get here?" he asked.

"I guess you could say there was a recruiter of sorts. Dave, Burke, and I agreed to come. There were a few others from the year above. Who wouldn't be interested in ghost-town living? We were told we'd come out here to tame the Wild West."

"Where's your family?"

"In Wilmington."

"North Carolina?"

Sailor nodded. "Or they were. Hurricanes, you know."

"My parents were in Cleveland. Years ago. The snowstorms." They were silent.

"How come there are no families here?" Cal asked.

"We believe in containment, you know that," Sailor replied. "We've got limited resources." He stood up straighter. "Plus, there isn't medicine if they got sick or enough food for them to eat . . . and what about providing them with an education?"

"But you'll die out."

"Us and the whole world." Sailor wasn't smiling. "The Land isn't against growth, Cal. We just choose who gets to join us."

"Oh, please. What about human nature? What about the desire to procreate?"

Sailor shrugged. "I'm not ready to be a parent, and in this world, I wouldn't want to be. Not ever. Micah says if we don't have examples of fatherhood to follow, we won't seek out that path. I think he's right."

"And the women?"

Sailor shrugged. "I'm not the person to ask."

Cal sat on the bed.

"I should go," Sailor said. "Leave you to your stuff."

"One more thing," Cal said. "Did the recruiter who came to Plank say anything about the Group?"

"I can't answer that question," he said, the color leaving his face.

"What do you mean?"

"The Group, it's not really part of the Land." He stepped backward. "Well, it is." He was almost out the door now. "But it's not that simple."

"Sailor, wait," Cal said. "What did the recruiter tell you?"

"Only what we wanted to hear." He paused. "That we had a purpose." And with that, Sailor turned around and was gone.

August had packed Cal's shorts and Frida's favorite blue dress. Their quilt, sewn by Frida's grandmother. A handful of Cal's bandannas. A few pairs of underwear. Cal didn't like to think of August going through their things, but thank goodness he had, because Cal had been craving another pair of socks and his second pair of boots and his pillow. They were all here.

August's ability to pick the right possessions felt like a seduction. It implied that he and the others could anticipate their needs, could guess what would bring them comfort and happiness. This delivery of possessions said, *We understand you.*

Maybe they did.

Cal pulled out Frida's blue cable-knit sweater and brought it to his face. The wool made his throat tickle. August had never seen Frida wear this, even Cal hadn't; she only used it as a pillow when she couldn't sleep. Somehow August had known she'd feel relieved, seeing it here.

It was the same as Micah knowing that Cal would feel flattered, honored, even, to be asked to join them each morning, to make important decisions. *You're really smart,* he'd said. *You're my brother-in-law.*

Micah wanted him to believe that their morning meetings were special but harmless. That they discussed who got what jobs and what repairs were needed right away and what members might be on the verge of dispute and need intervention. Village-elder-type work.

Their meetings probably did include such quotidian concerns, but there had to be more. Why else would he be asked to keep them a secret from Frida? How did the Land get all its coveted objects, for instance? Just yesterday, Frida had mentioned that she'd seen garlic powder in the kitchen, not due to expire for another year. And what was August looking for on his routes? The nature of his surveillance had to be central to their meetings.

If he joined them, he'd find out.

Cal swept his hand along the bottom of the bag and hit something hard. He didn't recognize it. He felt along its strange surface and felt wire, too. He pulled it out.

Frida's abacus. It was one of her artifacts, stored under the cots, where the turkey baster must have been hiding all along.

Why pack her abacus? If August had known of the child, Cal might have read its inclusion as a kind and hopeful gesture. Now it struck him as wrong, threatening even. It meant August had searched every corner of their house, that he'd left nothing unexamined. He must have been there for a few nights: like Goldilocks, he'd slept in their bed, eaten their food, tried everything on for size. He had seen the world as they did, or he'd tried to, though this wasn't about empathy but scrutiny and territory. He wanted Cal and Frida to know that.

Sailor had blanched when Cal asked him about the Group. No way this was an idyllic ghost-town kibbutz.

A place that banned children had to have a streak of insidiousness at its center.

These men were up to something.

13

In the dark, Frida held the sweater to her cheek. It itched and made her eyes water, her throat tighten, but maybe that was the point. She'd been nineteen when her father had given it to her, explaining it had magical properties. He was wearing it the night he met Frida's mother, and when the old Mercedes broke down for good on the 405. "This thing kept me from having a panic attack," he said as he held out the ratty sweater for her to take. Now that she was old enough to appreciate it, he said, he wanted her to have it. "You've got a job, right?" He'd rolled his shoulders back and forth, and shook out his neck, like he was getting ready for a boxing match. "Adult problems are just around the bend."

But tonight, the sweater wasn't working. Her eyes were open, and even her blood felt awake.

Cal was snoring next to her. He'd fallen asleep soon after telling her what had happened earlier that day, as if confessing had exhausted him. No wonder Micah and the others—Peter, August, even Sailor—had acted strangely at dinner. They knew about the baby. She and Cal might not be able to stay.

There were rules, apparently.

"And if we have to leave," Cal had whispered, before turning over in bed, "we'll never be able to come back." Frida tried not to hear the relief in his voice.

He told her he wasn't angry at her for lying to him, but she knew he was. At dinner he hadn't poured her a mug of tea as he usually did, and once it was just the two of them alone in their room he'd answered her questions curtly, hardly looking at her. He'd snuffed out the candle before she was even in bed.

She had apologized more than once because it didn't seem like he'd really forgiven her. He wasn't ready to accept what she'd done, and she got that. So she didn't push it. It was easier this way.

Cal thought she was sorry for misleading him—and she was—but she was also apologizing for the thoughts that had whiplashed through her mind as he explained what had happened in the Church. She didn't want the baby anymore. Just like that, she gave up that future. She was ashamed by how easily she let it go.

It's not even a baby yet, she told herself now. It was an embryo. It was a ball of cells.

She remembered something Toni had asked her, on one of their runs. Toni and Micah had been dating for four months by then. "When do you think life begins?" This would've been a weird question from anyone else, but not from Toni, who loved to muse and pontificate. Shallowness in conversation made her impatient. She had no use for small talk.

"There's a reason they call it small," she liked to say.

Toni had asked her this question as they jogged around the dirt track of the Silver Lake Reservoir. To get there meant a rough and sometimes dangerous bike ride, but it was worth it: it was one of the only tracks left in the city, and it was clean and wide, even if the reservoir itself was filled with debris instead of water. The homeless, rising in number, often used it as a toilet, and people said corpses were buried under all that trash: the rusted-out shopping carts and car parts, the gutted desktop computers, and the hundreds of plastic bags, porous with holes, swollen with brown rainwater, hanging from orphaned tree branches.

Cal didn't like her going, worried it was too dangerous, but she insisted it was fine. If Frida squinted her eyes toward the hills that overlooked the Reservoir, she could transport herself to a neighborhood that had once been beautiful, insufferably so, the wrecked houses above her transformed again to million-dollar bungalows of yore, painted in sage, avocado, pumice. She was good at editing the frame.

"Did you just ask me when life begins?" Frida remembered saying to Toni.

"Sure did," Toni said.

Frida could see her friend's tattered sneakers, hitting the dust of the track.

"I think it begins with consciousness," Toni continued. "The fetal brain really doesn't develop until the final months."

Frida hadn't had an opinion back then. But now, what did she think? She wanted to say she agreed with Toni.

"A person isn't a person until it can use its lungs," she had told Frida. "And those also don't develop until the final trimester."

"How do you know so much about this?"

Toni's voice was breathy from the run. "It's my job. In the Group. I read up on issues. I'm one of the researchers."

From Toni, Frida had learned a lot more about the Communities than she'd ever be able to discern from gossip sites. Toni was the one to tell her that Community members were encouraged to have one or two children, and if they wanted more, they needed a permit, which was pricey. "Because of the pull on resources," Toni explained. But if a couple couldn't have children at all, their status was threatened. "Calabasas, for instance, and Purell up north, really see parenting as the key role for every adult member of society," Toni said. "Some Communities are way more family focused than others, though."

"How do you know all this?" Frida had asked.

"I told you," Toni said. "I'm a researcher."

If Toni were here with her now, she might tell Frida that human life didn't begin until the baby was out of the womb, until it was breathing air. Whether that air was redolent with human feces and rot, or beautiful and pure, free of everything the city had burdened them with, didn't matter. Until the child was crying in the room with you, it was just a parasite in the female adult's body.

But, no. That kind of language was Micah's. Toni might agree with him, but her words, her cadence, would be different. She was gentle, and she had the gift of making Frida feel okay about being so pragmatic, so shrewd.

Toni would understand the calculation.

If Frida didn't care about this baby inside of her, if she could see it as something inhuman, then she might be able to rid herself of it. There were no children on the Land, but there had to have been accidents. They had to have access to the morning-after pill, at the very least. Or maybe there was some herbal remedy she could take—just something to make her bleed. She wouldn't think of it as anything but her period, come late.

I took care of it, she'd tell Cal. Wasn't that what women said?

She wanted to stay on the Land, and now they would be forced out. Back on the estate, she and Cal would become the Millers 2.0: starting a family in the woods, their kids hunting squirrels in loincloths, blissfully unaware of the world their parents had rejected.

Cal could fall in love with that life, but Frida knew how it turned out: some new settler would end up burying their bodies. For, surely, Sandy and Bo had eaten that poison because they'd finally faced despair head-on.

Cal twitched in his sleep. If she were to tell him what she was thinking, he'd be angry, afraid, worried. Already, he loved their child. It probably had no eyes or limbs, no intestines, but already there might be a heartbeat. *Politics aside,* she imagined him saying, *that's where life begins. My child's, at least.*

And, well, shit. She knew she was wide awake because she agreed with him.

Frida must have fallen asleep somehow, because when Anika knocked lightly at the door, whispering that she'd be waiting downstairs, Cal had to shake her awake.

"Bread," he croaked, and turned over. The room was as dark as it had been at midnight, not even a bird cutting the silence. How had Anika woken herself up? Cal thought certain people on the Land had alarm clocks, though if that were true, he and Frida would have heard them by now. Frida joked that Anika probably slept with one eye open, ready to grab a weapon and fight off an intruder, never really surrendering to dreams.

At fourteen, Micah had become interested in Sparta. He downloaded a bunch of books on the subject and had even emailed a professor at USC for more information. Frida remembered him telling her about the military training for Spartan boys: how they were sent in groups into the wilderness with just knives to fend off wild animals. Perhaps this was how Anika, fierce protector of the larder, had been raised. She *did* seem tough. Micah had probably intuited Anika's strength and promoted her himself. *Anika,* she imagined her brother saying, *I pronounce you Leader of the Cooks, Protector of the Knives, Keeper of the Fire.*

When Frida got to the kitchen, Anika was wiping down the center table with a rag. The room was lit with two large candles, and in the weak light, wise and solid Anika looked like a sweet old woman in a storybook, the kind who might lead lost little girls and boys back to her cabin for a warm meal, offering to dry their wet socks by the fire.

She glanced at Frida. "You took your time," she said.

Frida was surprised Anika didn't whisper.

"It was so dark, I had to feel my way down the hallway with my hands on the walls." Frida smiled, even though Anika remained serious. "I took those stairs *very* carefully."

"I forgot what it might be like for you, being new. This place is so familiar to me, I could cook in here with my eyes closed."

"I don't doubt it."

Anika put down the rag. "This isn't when I usually make the bread," she said.

"It isn't? Burke said you're in here baking practically every morning."

"More like every night. I'm down here much earlier. Otherwise, there's no time."

"You could do it in the afternoon."

"I don't like people around."

"No?"

She shook her head. "I have trouble sleeping as it is, so night is preferable."

"I'm sorry to mess up the schedule."

Anika waved her off. "It's fine."

Frida's eye caught a pile of corncobs at the end of the table, stripped of their kernels, as square edged as honeycombs. Next to the pile was a glass bowl filled with the white and yellow kernels, strands of corn silk stuck between them like food in teeth.

"For cornbread?" Frida asked, gesturing to Anika's work.

"That's just prep. We're having chowder tonight."

"You sure do love soup, don't you?"

Anika raised an eyebrow, and Frida knew she'd said the wrong thing. She couldn't let Anika think her ungrateful or picky.

"We're not making bread this morning, just so you know."

"We aren't?" Frida asked.

"I know you'd love to, but that's something you have to earn." Frida wasn't sure how to reply.

"I'm not going to give you bread duty out of pity or obligation."

"Who would you be obligated to? My brother, you mean?"

"Nepotism has never been a problem around here," Anika said. Frida waited for her to say, *Until now,* but she didn't.

"Why am I awake then?" Frida asked. She tried to keep her voice low, to control its quiver, but she knew it was giving her away.

"Relax," Anika said. "I want us to bake a cake. You know how to do that, don't you?"

"Of course," Frida said, swallowing a tide of spit. "I love baking cakes."

It was true, though she hadn't baked one in years. In L.A., she and Cal didn't have money to waste on things like baking powder and sugar, and then the electricity stopped working, which meant their oven became a glorified cupboard to store the extra candles they used, sparingly, at night. Soon after they met, August had tried to get her to trade for eggs and flour, but she'd have been nuts to go for it; he'd wanted their coffee and Cal's heavy coat. Even though she would've only been able to batter fish with the flour, she sometimes regretted turning August down and imagined all that she might do with it. She could feel its dust in her lungs.

"What's wrong?" Anika said. "You look like you might faint."

"You have everything to bake a cake?"

"*Cake* is probably the wrong word. We have to use this oven, so I want to do something simple. It's more of a clafoutis."

Frida could have laughed. She hadn't heard that word in a long while.

"Think of it as a sweet pancake," Anika explained. "It's French, and traditionally made with cherries."

"I know what it is, Anika."

"Do you now?" Anika shrugged, as if to say, *You cannot impress me.* "I need to get the baking crate from the cellar. Wait here."

Baking crate. She'd said it as if this were a thing that people all over the world had.

A moment later, Frida was alone, the two candle flames emitting an uneven, wobbly glow across the kitchen. She had spent enough time in this room that she didn't need to rely on the sunlight to arrange its details. There was the hand-scrawled sign above the dishwashing trough that read daydreaming wastes time! and the umbrella stand next to the woodstove held barbecue tongs, shovels, and metal tools to stoke the fire. The two cellar doors, made of beautiful pinewood, and the big black smoke stain on the ceiling. On the windowsill, a line of pumpkin seeds; some benign troublemaker had pushed them into the most recent coat of paint before it had dried. It amazed Frida that not one seed had been pried off.

"Taking it all in?"

Anika stepped up from the cellar with the baking crate on her hip. She shut the doors with one hand, just as she'd done when Frida had first met her. They closed with a definitive *thwap.*

"I guess I'm trying to figure this place out."

Anika hardened her eyes in that combative way she had and put down the crate. Before she began pulling anything out of it, she grabbed for a metal bowl that had been on the table all along. It was covered with a dingy white dishtowel, and Frida hadn't noticed it.

Anika removed the towel to reveal a pile of brown eggs, speckled with shit. "Have you met the chickens yet?"

Frida hadn't, but she'd heard about them from Sailor. He told her about the first time he was assigned to butchery for Morning Labor. He'd been afraid he would vomit at the sight of the dead animals hanging from a post behind the barn, draining blood, but he'd signed up anyway, because he wanted to challenge himself. What happened next surprised him. He was stunned by the beauty and simplicity of the process, he said. "Invigorated even." Sometimes it felt to Frida as though Sailor were chatting her up at a bar, telling her whatever stories would keep her listening, make her suck down her liquor faster. Other times, she was less cynical; it was just that he'd truly welcomed her to this place, to their life out here.

"My favorite is Suzanne," Anika said, not waiting for Frida to answer. "Her eggs are divine." She reached into the baking crate and pulled out a series of Mason jars, each of them filled at least halfway and labeled with masking tape and thick black pen: flour, sugar, baking powder, baking soda, salt. If Cal saw all this, he'd flip.

Before asking her next question, the one Cal would want her to ask, Frida steeled herself, like she used to do before running across a four-lane boulevard, the break in traffic impossible to measure, dangerously unpredictable.

"Where did these come from?" she asked.

"You just saw me go into the root cellar."

"You know that's not what I mean."

"What *do* you mean, then?" Anika asked. She paused and reached for a brown container at the bottom of the crate, its label still intact. Frida didn't need to see the other side to read what it said: Hershey's. It was cocoa powder. It was chocolate.

"Oh my God," she whispered. She could already smell its rich, slightly chalky scent.

"We have a saying here," Anika said as she placed the container on the table. "Don't get involved if you're not ready."

Frida wasn't sure what Anika meant but couldn't take her eyes off the brown container with its plastic top, its nutritional information printed along its side in black and white, and the big bold letters across the front. Back in L.A., chocolate, even the mass-produced kind, cost more than a week's wages. If you lived in a Community, you could get it easily; that's what Toni had told her once. She said some were still producing it behind those impenetrable walls.

Frida felt suddenly nervous. Anika was unpacking the baking crate with care. She was putting this all on display for Frida. But why?

"I want you to do everything," Anika said. She handed her a set of nesting bowls, a measuring cup, and the measuring spoons.

Frida nodded, aware that this little cake-making party was a test, part 1 of Anika's baking exam. It was also show-and-tell. If this morning Frida saw the Land's cocoa and flour and sugar, then what might tomorrow bring?

Frida grabbed the largest bowl and began measuring out the sugar. She knew exactly how much to use. A clafoutis was easy, especially if Anika had an understanding of the stove and its tendencies.

Anika stepped out the back door and returned with one of the glass pitchers of milk. "I asked Lupe to milk Jessa last night and leave some for us."

"I know there aren't any cherries," Frida said, "but do you have any fruit at all?"

"Some apples, I think. Why?"

"I'd like to use them. Skip the chocolate. It doesn't fit." She met Anika's eyes, and for once, the woman looked away.

After Anika had retrieved the leftover apples, she stepped back from the table and watched Frida with provocation in her eyes. She was daring her to mess up.

As Frida cracked the eggs, poured the milk, and sliced the apples thinly, she pushed Anika and her judgment out of her mind. Forget her. Frida could, and would, enjoy herself.

"Don't you have to get the oven going?" she asked. She tried to hide the smugness spreading across her face when Anika was forced to turn away and fulfill her duties.

Frida poured the batter into the round pie tins. She had always loved baking for the time it took, for the patience it required, and dexterity, too, if you wanted your results to be beautiful. It was about risk as well as precision: you never knew if a dessert was good until your guests were taking their first bites.

But Frida didn't care if these cakes turned out badly. She wasn't vying for head pastry chef. If she failed at this ghetto clafoutis, Anika might let her try baking bread because she wanted another good laugh. And if Frida made something delicious, Anika would be too busy eating every last crumb to say something snarky. Frida felt her bravery rise. She felt emboldened.

She handed Anika the tins to place in the oven. "Did you know the Millers?"

Anika held her face perfectly still, as if she hadn't heard.

"Anika?"

"Who?" She turned to the oven, a small fire going inside of it.

Frida couldn't help but laugh. "You're a terrible actress, you know that?" She waited for Anika to respond, and when she didn't, Frida decided to take a risk. "Sandy didn't like red, and neither do you. You gave yourself away when I cut myself."

Anika spun around, and Frida thought she was smiling. But, no, she was grimacing, and her missing tooth made her look witchy or homeless, or both: a sorceress who slept beneath an overpass and shit in the Silver Lake Reservoir.

"I hoped you'd think I was just being squeamish," she said. She had turned back to the stove again and was pushing the tins onto a metal grate that sat atop the small flame.

"The color makes you nervous. Sandy was the same way." Frida paused, and she sensed her own brother in her voice, guiding its tone.

Anika straightened her posture and looked back at Frida.

"Tell me what it means," Frida said.

"You mean you don't already know?"

Frida wasn't sure what else to say. She didn't want Anika to know that she'd already talked to her brother about it, or had tried to. "I saw Sandy freak out about the color only twice. But I remember both times, because it seemed strange. She was frightened, like you were."

Anika let out a tiny mewl, then stopped suddenly, as if embarrassed. "She was my friend."

"Mine, too."

For a moment they both watched the oven. Frida hoped they would be able to smell the cakes over the smoke of the fire. How would they cook in here without burning everything?

"I can't believe they're dead," Anika said after a moment. "It's easier, sometimes, to think of them as just a ways off, living separately."

"I'd do the same, I think."

"We were both here from the beginning," Anika said, and

225

Frida thought of the phrase Micah had used in the tree house: *original settlers.*

If Frida was silent, maybe Anika would say more. She practiced a trick Micah had taught her when they were in high school. She silently counted backward from ten.

At the number five (it was always five), Anika began talking. "It started because of the Pirates."

"I still can't believe they're real. That they exist."

Anika laughed meanly. "Of course they do. How lucky for you, to be able to think otherwise."

"I would've thought you were protected, by the . . . Forms."

"We didn't always have so many surrounding us. Most were built later."

Anika's eyes were back on the oven.

10-9-8-7—

"Soon after we arrived on the Land, just a week or so, two of our men were killed. When we found them, they were naked, their bodies . . . they'd been mutilated, sliced up. They were covered in blood, just covered in it."

Anika spoke as if she had never told this story before. Hilda used to call that kind of story a slumber-party confession: the teller experiencing shame and relief in equal parts.

"After that, for a while, there was nothing. We went about our business, building shelters, getting the garden started. And then, one day, I found a red rag tied to a wooden stake, shoved into the dirt right outside the Hotel. I didn't know what it meant, and no one had seen the rag before, let alone the stick." Her voice went quiet, and Frida had to lean forward to hear what she said next. "The next day, the Pirates came back."

Frida held her breath.

"There were probably thirty of them. They were all youngish men, and, I can't explain it —the *greed* in their eyes. It was like

226

they were just sitting down to a big feast." She paused, shaking her head. "We outnumbered them, but over half of us were female, and we were vulnerable, and scared. None of us, the men included, had experience fighting. In our past lives, we'd been scrap-metal collectors, soup-kitchen coordinators. What did we know?

"We had only a few guns, and we were running out of ammo, and they came in on horses. They rode up slowly. I remember the sound of the horses trotting across the dirt as they approached, and how we came out to see who it was. They were wielding guns. Some had knives. We'd been naïve to think we just had guests."

Frida shivered, just as she had on the ride out of L.A., thinking about men hurting her and Cal. On the drive, she'd made herself so cold with worry she had to be covered with a blanket at all times. Now she crossed her arms as Anika continued talking.

"That first time, half of them dismounted, coming toward us while the others just lingered. I remember the smell of those horses. The Pirates themselves were covered in sweat and dirt, thick as a second skin. They wanted food, they said. They took our guns. They made John give them his boots. Four of the women ... they were—"

"You don't have to say it," Frida whispered.

"I didn't see anything. I have nothing to tell you. They took the women into one of the half-collapsing houses, and the rest of us waited, guns to our heads."

Anika was looking at her, as if waiting for her to say something, but Frida didn't know what. Nothing would be sufficient. The smell of the clafoutis had started to fill the room: sweet and warm, a comfort if there was still such a thing.

"Later we tore down that house," Anika said finally. "We

didn't do it completely, we left pieces of it up. As a reminder, I guess, that we had survived. The brick wall that Cal's team is dismantling? That's it." She stopped. "We tried to give the Pirates all of our vegetables. We didn't care if we starved, we just wanted them to leave. But the man in charge—he had long hair, and eczema or psoriasis all over his arms, he was scaly like a snake—he took only half of what we had. He would be back. He said he needed us alive to grow more.

"The men wore red. Bandannas on their heads, red shirts if they had them. And their hands, their fingernails, they seemed stained with it. Bloody." Anika turned to her, condescension spreading across her face. "You see, Frida, red became the color of violence." She spoke as if she were a teacher, reciting to a particularly dense student a lesson she had been explaining for days. "Every time the Pirates were coming, they'd warn us with something red, usually a piece of fabric, but once it was a red-handled shovel. Another time, red paint splashed across the side of the Church. They wanted to get inside our heads. You'd think the warnings would help us prepare, defend ourselves, but they just got under our skin and made things worse. That's exactly what they wanted.

"Once we tried to make a plan. As soon as we were warned by a red object, the four women hid in one of the smaller houses—there was a crawl space that was hidden well. We knew they couldn't handle seeing those men again, they wouldn't survive, and so we made them as safe as we could. The rest of us hid in the Church. We figured we'd be enough for those monsters—as a group we'd distract them. We barricaded the doors, waited with our last scythe, the one they'd left us to garden with. We heard them outside, tying up their horses, laughing, calling commands at one another. They banged on the door, but didn't try to get inside."

"They didn't? They just left?"

Anika shook her head. "They were waiting us out, and it worked. By the fifth day, we unlocked the doors. One of our men was very sick, he needed water, and all of us were starving. We needed food. We'd been shitting in buckets. If one of those Pirates wanted to shoot me, I would have welcomed it. I really thought they were going to, too. But they just burned down the barn we'd recently built and took off with our reserves of grain."

"They didn't hurt anyone?"

"Not anyone in the Church."

"The women," Frida said. She wanted to reach out to touch Anika, but she was afraid Anika would flinch.

"Once the Pirates had left," Anika said, "we went to the crawl space. It was busted wide open. The women were gone. We'd assumed they wouldn't be found there, that they'd remain hidden. They had felt almost safe, tucked away like that, but they weren't. We'd been so stupid to hide them there, alone. When we found the crawl space empty, we thought they'd been kidnapped, but two days later, we found their bodies in the woods."

"Oh, Anika."

"You want to know why we didn't leave. You think we were asking for it."

"I don't."

"Where would we go?" She sighed. "The point is it didn't take long for the color to turn my stomach. As stupid as it might sound, red scared all of us. Even something left out by accident—if it was red, we panicked."

"It doesn't sound stupid at all."

Anika grunted. "Finally, when Pilar had a breakdown at the creek, sobbing as she tried to wash a red dress, we decided to destroy everything that color. Just rid ourselves of it completely."

229

Anika smiled, but it was woeful. "Believe it or not, it made me feel better, temporarily at least."

"I bet." It sounded so dumb, but she didn't know what else to say. Frida remembered the way Anika had looked away from her cut. She hadn't described later attacks —and there had to have been more. Had she seen a Pirate's hands up close? Had they touched her?

"I've tried to shake the red thing," Anika continued, "but it's hard. Sandy once said it was like rejecting religion. We did that, long ago, it was partly what united us. But she said it was like turning your back on God and then catching yourself praying every now and again."

"That sounds like something she might say."

"Anyway," Anika said, and moved to the table. She began putting the lids back on the jars and returning them to the crate. "It's over. The Pirates are gone."

"But how? How did you get rid of them?"

Anika had the cocoa tin in her hand, and she raised it like a judge's mallet. "Your brother, Frida. He's the one who helped us. When Micah and the others arrived, we were able to keep the Pirates away. He protected us." She put the cocoa into the crate.

"Did he fight them off?"

She sighed. "He came with guns, and more men, strong ones, who wouldn't be intimidated. He taught us how to protect our land."

"And then he came to live with you guys here. I guess it was a smooth transition."

"We owed him," she said, "for what he did for us."

The sun was rising. Morning Labor would begin soon, the crew arriving any second now, yawning, rubbing their hands together for warmth.

Frida took the rag Anika had been using and dragged it across the table. But then Anika's hand was atop her own, as if to stop her from cleaning.

"Little Janie," Anika said. This time, she whispered.

Frida looked up. Anika's eyes revealed nothing.

"Jane? Jane Miller? You knew her?"

"By the time you met her, did she talk okay?"

"You mean you—"

"Good morning, sweet ladies." It was Burke.

Anika stepped away from Frida. "Burke, leave the kitchen and come back in with a more appropriate greeting."

Without complaint, Burke turned and left the room. Frida squeezed the rag in her fist.

"You knew Sandy's kids?" Frida whispered. "Were they born here?"

"'*They*'?" Anika said. She looked horrified.

"Anika? What's wrong?"

But already the rest of the crew had entered the kitchen, and the two women were separated as currents in a wave.

14

Word of Frida's baking traveled as fast as gossip. Just an hour into Morning Labor, and already three people had told Cal about Frida's "sweet pancake," which sounded to him like the name of an unfortunate and sparsely attended burlesque act. Everyone had urged him to run over to the kitchen to try the cake before it was gone, but he didn't. He had work to do.

Of course everyone was smitten with Frida and her talents. Cal could understand it, but that didn't make him feel better. It had been years since Frida had baked anything; in fact, he was sure the last time had been for his twenty-fourth birthday. By necessity, and because they lacked funds, she'd baked him a vegan sugar-free cake, sans icing. It looked like a waterlogged block of wood, and Frida had cried as they ate it. It had tasted okay.

Now, the only thing that comforted Cal was that Frida hadn't gotten to bake bread this morning. She was probably wrestling with the disappointment, and though that saddened him, made him feel vaguely protective of her, just as he'd felt when she served him that birthday cake so long ago, the feeling didn't overshadow the pettiness in his heart.

The thing was, Cal had woken up happy. There was no longer a knot of secrets between him and Frida. Micah knew the situation, and the choice to stay on the Land was out of their hands. Frida's body would begin changing soon, and there would be something to look forward to, no matter what happened here. Cal knew he was being naïve, stupid even. But he didn't care. All his life he'd been careful, hesitant. Now he'd have what he wanted, and what he wanted was a family.

He'd caught himself whistling on the way to Morning Labor. He hadn't gone to the Church for the private meeting; he figured the invitation had been rescinded since he'd revealed Frida's pregnancy. Besides, he had an interest in the work he'd been doing with his team. The outdoor oven was nearly done, and looking at it, he felt a surge of pride. It was well made, almost elegant. He didn't even chastise himself for feeling in tune with his comrades, for thinking of them as comrades, for using phrases like *in tune*. It wasn't until he heard about Frida's baking that his day had turned south.

He finally forced himself to forget about it and try to focus on the oven. There was still work to do, and he'd do it well. It amazed him how much satisfaction there could be in that.

Once he'd settled into a rhythm, his back aching, the paste for the brick mortar clinging to his skin, Peter showed up.

"Can you come look at the garden irrigation system with me?" He spoke quietly, his gaze straight ahead, as though they were two undercover agents. "Your brother-in-law says you're a gardening genius, that you have professional experience."

Cal held up his hands, as if to say, *Hey, sorry, man, I'm busy.*

Peter took a rag out of his back pocket and offered it to Cal.

"It wipes off easily," he said, and waited as Cal cleaned his hands.

It was strange how none of his comrades said anything as Cal left with Peter, not even Sheryl, who was normally such a pill about the rules. In fact, no one looked up as the two men walked away. Maybe he and Peter were undercover after all.

As they headed in the direction of the garden in silence, Cal felt as though he was in trouble and was being led to the gallows. Would Frida be there, too? He pictured Micah holding two ropes and the requisite black hoods. No pillowcases here. Cal wasn't about to underestimate what the Land had access to.

He was being dramatic. If Micah had told Peter about Cal's experience running gardens back in L.A., it meant they needed that kind of expertise, that they needed Cal. He wasn't beholden to them; he had something to offer, too.

When they passed the garden, Peter not even slowing down, the dread that had been collecting at the bottom of Cal's spine spilled down his legs.

"I thought we were going to the garden," Cal said.

"Later," Peter replied walking more briskly now. He was heading toward the woods, and Cal could do nothing but follow.

"Where are we going, then?"

"Micah wants to see you."

"And you do his bidding?"

Peter didn't respond. He couldn't be goaded into anything, Cal realized. Peter was too mature to be embarrassed, too powerful to worry about what the new guy thought of him. He'd probably exuded this since childhood; he was a natural leader.

"Micah and I both want to talk to you," Peter said, and that was it.

He led Cal into untamed forest at the northern edge of the Land. Cal and Frida had come from the west, and they hadn't had a chance to explore the rest of the borders. Cal had seen this section of woods from afar and wondered about them. The Spikes rose on either side, waiting like armed guards, and he imagined there was a whole maze of them deeper in the forest.

"Can I take a look at the Forms sometime?" Cal asked.

"I suppose so," Peter said, and pushed aside a mass of thorny branches. He gestured for Cal to walk ahead. "You'd probably be good at security. I can tell you've got that kind of mind: you're the paranoid sort. Always assuming danger."

Cal followed Peter's lead and stepped over a rotting log. There was a path here, but it was tricky. "I want to know how the Forms are really a threat to outsiders," he said. "I mean, come on."

"They scared you, didn't they?"

Peter kept walking, going around another rotted log and pushing aside tree branches. He stepped over what looked like a dead bird, covered in flies. "Watch out," he called back, and Cal stepped over it, too, holding his breath.

Finally, Peter stopped at the trunk of an Oregon oak. He put his palm against it.

"Where are we?"

Peter pointed up, and Cal saw that there was a wooden platform built into it.

"A tree house? How quaint."

A big laugh sounded from above. Micah. "Come on up!" he called.

"Go ahead," Peter said when Cal looked at him.

Cal shinnied up the trunk without using the footholds. There'd been a plane tree on his father's farm, great for climbing, and as a boy Cal had loved to hang upside down from its highest branch until he felt the skin of his face turn purple.

"Look at you, Tom Sawyer," Micah said when Cal pulled himself onto the platform. He was sitting cross-legged on the floor, hands loose on his knees, as if he'd been meditating.

"You're in a good mood," Cal said. Peter was coming up behind him.

Even though there were two camping chairs at the edge of

the platform, Cal sat on the floor in front of Micah. Peter did the same.

"Does this mean I'm still in the cabal?" Cal said.

"'Cabal'?"

"He means the meetings," Peter said.

"I know what he means, Peter." Micah had his eyes on Cal. "Do you see August here? Or even little itty-bitty Sailor?"

"Why pull me away from Morning Labor, then? We're putting the finishing touches on the outdoor oven. I should be there."

"Already so committed," Peter said. To Micah he added, "I told you."

"Told him what?"

"Jeez, Cal," Micah said. "Take the venom out of your voice."

Peter nodded. "All I said was that you're good for this place."

"Am I?" Cal flung his legs in front of him. "Are you ready to be an uncle, Micah?"

Cal wasn't sure why he was being so cavalier.

"That's why you're here," Micah said icily. "To discuss the matter."

"I guess I've given you a lot to think about," Cal said.

Micah leaned back on his hands and hung his head back so that all Cal could see of his face was his beard. A few crumbs were stuck there like flies in a spider's web.

"Do you like the tree house as much as Frida did?" Micah asked.

"She's been up here?"

Cal immediately wished he hadn't said it. This was where she must have talked to Micah. Why had she left out that detail? His face felt hot; he might as well have been hanging upside down from one of the tree branches. Damn it, Frida. He thought they were done with secrets. Soon everyone would think he didn't know his wife. Maybe they'd be right.

Micah lifted his head up. He was smiling. "My sister sure is secretive, isn't she?"

Peter rolled his eyes. "Guys, knock it off."

"Knock off what?" Micah said. "I guess your news has stayed with me, California."

"As I suspected."

"For one, how are you sure that Frida's *with child?*"

"You know how the female body works," Cal said. "She's late."

"That doesn't confirm a pregnancy," Micah said. "Not these days."

"She might have missed it for a number of reasons," Peter said. "Poor nutrition, for one. Micah says she used to be heavier."

"She's lost some weight over the years, yeah," Cal said. "The grocery stores in L.A. weren't exactly well stocked by the end. And out here, just the two of us, it's not easy."

"There's also early menopause," Peter said. "It's been known to happen."

"Stop it," Cal said. "Look, guys, Frida says she's pregnant. She says she's sure, okay?"

Micah laughed. "Fuck me! Frida, feeling the pull of the moon? We can't be talking about the same person here, Cal. My sister used to throw out a pair of panties every month because her period always, as she put it, *surprised* her."

"Jesus Christ, Micah." Cal didn't know what was worse: Micah talking about Frida's body, or that he was right. When she and Cal had first started dating, Frida had to buy new underwear on a regular basis. "Oops," she'd like to say, coming out of the bathroom.

Later, when the department stores went out of business, and they lost their Internet connection for good, and they had hardly a dollar to spare, especially on clothing, Frida committed herself to being a little more "organized." That's when she realized she had a perfectly predictable cycle. "I'm textbook," she'd cried, delighted.

Before then, Frida's relationship to her own body had puzzled Cal. It was funny, even charming, how ignorant she was of it. But from another angle, it seemed pitiful. Or just weird: how could she not be obsessed with a body like hers? In the beginning, Cal had thought of it all the time. He remembered one time at work in L.A., planting tomatoes and thinking of Frida's smooth back and her pillowy ass, which he loved to spread apart.

Peter cleared his throat, and Cal realized no one had said anything for a moment.

"Micah," Peter said, "that really is repugnant."

"What?"

"I had a sister."

Past tense, Cal noted.

"And I stayed far away from her, and her . . . period. It's weird to talk about it."

"Don't be a child, Peter. You misunderstand me. You both do." He turned to Cal. "I bet all these years, you thought Frida was just being absentminded about her body."

Cal didn't reply.

"You're wrong. It has nothing to do with her period, or her womanhood, or some shit like that. It's time she doesn't get. If Frida doesn't keep track of time passing, then it can't pass. Then nothing changes."

"Well, she keeps track now."

A gust of wind picked up, and from miles away, a scrub jay cried out. The platform creaked beneath them, and Cal imagined the whole thing toppling to the ground. No one said anything. The tree swayed.

Peter stood, maybe to break the mood. He leaned against the railing. "Let's stay on topic, guys. Cal, has Frida been exhibiting any symptoms?"

Micah laughed again. "'Exhibiting.' Nice one, doctor."

Peter shot Cal a look that meant *Don't mind him, he's just showing off.* Cal couldn't help but feel relieved; here was an ally.

"Is she nauseated?" Peter asked. "Tired?"

Cal shook his head. "Not that I know of. But it could be too early."

"There you have it," Micah said. "There's nothing definitive."

"But—"

Peter held up a hand. "You know he's right, Cal. And if he is, then things are a lot less complicated."

"I realize her pregnancy is a complication for you," Cal said. "But not for me."

"We both know that's not true," Micah said. "Frida wants to stay, and I think you do, too, if you'd just allow yourself to admit it." He'd moved to a kneeling position and was fiddling with a nail in one of the wood planks. He would not look at Cal. "Tell you what. Let's keep this supposed pregnancy quiet."

Peter was nodding.

"What about the Vote? It's supposed to happen next week."

"Since nothing is conclusive," Micah said, "we want to keep it out of the decision."

"But what happens if it is conclusive? What if she's pregnant?"

Peter said nothing.

Micah continued to fiddle with the nail, his fingers poised claw-like, and he bit his lower lip in concentration.

"Micah? I asked you a question."

He looked up. "There are options."

"What does that mean? Do you guys have an abortion clinic

set up here? Or do you send women to a Community to take care of it?"

Peter sighed. "Most Communities don't allow abortions. Didn't you know that? Those fucking Christians."

"So what if Frida's pregnant? What then? Will you guys just bring out the wire hangers?"

"Stop it," Peter said. "You're being paranoid, once again."

"*Options* is a generic term," Micah said. "Your response is a bit of a Rorschach test, no?"

"Fuck off, Micah. Why don't you just tell me the whole story? Why aren't kids allowed here? Sailor told me the party line about containment. But that doesn't really explain it."

"It doesn't?" Micah said. "It seems perfectly logical to me."

"It's a long story," Peter said.

"So tell it. Where else do I have to be?"

"The Land has to stay contained, Cal," Micah said, "so that word of this outpost doesn't grow. I'm supposed to be dead, remember?"

"Forget about that," Peter said. "Have you ever thought about how hard it would be to raise a kid out here?"

Cal heard something sorrowful in Peter's voice, and he thought of Jane and Garrett. Burying them. He stood, and his knees cracked. It sounded like a branch breaking.

"Listen, Cal," Peter said. "This is in your best interest. We want you to work with us. And if Frida is pregnant, then we'll discuss it when the time comes. I'm just not convinced she is."

"But why risk it?"

"Because," Micah said. "There are options."

"We want the baby."

No one said anything. Why did Cal feel like Micah didn't believe him?

"I still don't get why you want me and Frida here. It seems like all we're doing is causing you all a lot of trouble."

Peter smiled at Micah. "That's what I kept asking."

"'Kept'? Why did you stop?"

Peter nodded at Micah.

"My sister," Micah said. "She's here."

"So you're human after all," Cal said. "Is that it?"

Peter actually laughed. It was such a clear, pure thing. Cal could see the man respected him. "This is why we need you in our morning meetings! To put Mikey in his place."

Micah practically growled. "I realize you and Frida are a package deal, whether I like it or not. If you're here, we might as well use that noggin of yours. Our garden isn't doing well— the irrigation system is clogged. Go make yourself useful."

Meeting adjourned. Cal used the footholds on the way down, and Peter told him he could find his own way back. Cal was flattered that Peter had that much confidence in his sense of direction. He was also smart enough to know that Peter was aware of this.

Before Cal left the woods, Peter told him to talk to Frida himself, and as soon as he could. "She should have no problem with keeping the secret," he'd added, and smirked. Cal wanted to spit in his face for that, even though it seemed like Peter had been on his side in the tree house. He was probably the only reason Cal hadn't pushed Micah out of the tree. And, anyway, Peter was right: Frida was having fun with her little secrets. Hopefully, for once, she would do what Cal asked and keep her mouth shut. She could be so selfish sometimes. "I'll see you tomorrow," Peter said. "At our meeting." He smiled. "You're in the cabal, Mr. Paranoid. Get ready."

There it was. Tomorrow morning he would head to the

Church along with the influential men. The move would be noted by everyone else on the Land. After that, wasn't voting just a formality? If the others wanted him and Frida to leave, wouldn't Micah step in to veto their decision?

He would go to the meeting tomorrow because he wanted to understand how the machine worked. Micah had intuited that immediately. He'd give Cal what he wanted, but Cal would have to pay for it. Cal just had to figure out the price.

Working in the garden, Cal felt the foul mood that had threatened to take over all morning crouching in again. His hands were muddy, a blade of grass had dug itself into his thumbnail, and it seemed the woman he was working with, Rachel, knew next to nothing about . . . well, about anything, really. But that wasn't any excuse for his being so rude. He was sighing like a sullen teenager every few minutes. Rachel didn't deserve his crabbiness; after all, she had to sit in the dirt, too. At least the canal would be cleared soon, and then she could go sit at a table with someone more pleasant.

"That does it," he said, and they both got to their feet.

"Thanks, Calvin," she said.

"Can I ask you a question?" he asked.

"Sure."

"Are you going to vote?"

She laughed. "Oh boy."

"It's none of my business, sorry."

"It will be soon enough. The Vote is public. Didn't anyone tell you that? You'll see who wants you to stay."

"And who doesn't."

"I wouldn't worry."

"You don't even know what outcome I'm hoping for."

"Oh, please." Rachel raised an eyebrow. Cal thought he could see an old piercing there, the tiny hole abandoned; it had been left to close, but it was stubborn, it wouldn't.

"Everyone knows you're tight with Micah and them," she said. "You're in the meetings now, aren't you?"

Cal was too stunned to answer. How did she know?

"I've gotta get some food in me," Rachel said. "A sweet pancake maybe." With that, she turned and left Cal alone in the mud.

Cal went to find Frida to tell her what Micah and Peter wanted. They were right; Frida couldn't whisper news of her pregnancy to a soul. "I'm serious," he'd say. He wanted to protect her, but he couldn't say that: she'd laugh or, worse, be offended. She'd tell him she didn't need his help, his strength, his useless male bravado. For all he knew, Frida was keeping secrets just to prove he couldn't keep her safe. But this wasn't a final paper in a women's studies class; Frida needed him. He needed her, too.

If Frida could display a few symptoms, that would shut Micah and Peter up. Cal believed what he'd told them, that it was still too early in the pregnancy, but he couldn't help but wish a little nausea on her, a fatigue that dragged her into a nap every other hour. When he went to find her in the Hotel dining room, Anika had said she was resting, tired after waking up so early to bake. Cal's heart sped up. Maybe it had nothing to do with getting up early, maybe this was the beginning of the symptoms he had been hoping for. He'd ask her when they were alone.

It was funny to think that way now: *when they were alone.* Before, that's all they ever were. He'd loved being the only two people around for miles; he understood that now. The life they'd created for themselves had been fragile and solid at once, and beautiful in those ways, too: the shell of an egg, the stone of a pillar. Now things felt wrong. These people had no idea what Frida was like, what she needed, what she called out for in the middle of the night when she was afraid, when her stomach hurt, when she just wanted dawn to come and ease the dark. She and Cal had been through so much. It was like Frida didn't agree, like she didn't care.

Cal found Frida in the outdoor lounge with Sailor and Dave.

Dave had shaved, and without the scruff of hair covering his face he looked younger than before, and better looking; his beard had been hiding a strong jawline and a wide smile that made him look almost arrogant.

Did he want to impress Frida? Dave had been so rude when they'd arrived—the suspicious glances, the rough way he'd handled their things—but maybe by now he'd cooled off. And Frida was the new girl in town. The first and last, supposedly.

When Cal was a little boy, his mother had told him that someday his true love would seem different from everyone else in the world. "Like a bright red car in a sea of jalopies," she'd said. It struck him that, although his mother had not been in love with his father, nor with her various long-term boyfriends, she'd been right. This was exactly how Cal felt, looking at his wife. His red car.

Dave saw Cal first. He waved, and Frida turned.

Though she smiled and called his name, Cal thought he detected a microsecond of disappointment on her face. It reminded him of how she used to act after spending the day with Micah. It was as if she'd gotten so used to her brother's inflections and cynicism, and the way he could make her laugh, that returning to Cal jarred her. He wondered sometimes if Micah made fun of him to his sister, so that when Frida saw Cal again, she had the urge to laugh and had to force herself not to.

"Hi, babe," he said now, and bent down to kiss her on the mouth.

"Hi," she said.

What was it that had fled so suddenly from her face? Was it that she'd been sitting with two attentive men she didn't know very well, their lives mysteries she could mine for years, and Cal had barged in to interrupt the moment? She'd been so happy just seconds ago, as giddy as she'd been when she first met Sandy Miller by the creek. But now that Cal was here, breaking up playtime, she looked, if not unhappy, then concerned. Perhaps she was worried about what he'd say.

"Nice face," Cal said to Dave, and Frida giggled.

"He's a looker, isn't he?" she said, her hands clasped under her chin like a cartoon animal in love. "Sailor's jealous."

"Am not," he said, pouting.

"Poor baby," Frida said. She was laughing again.

"I heard about your fancy pancake," Cal said.

"It was amazing!" Sailor said, and Frida fake-protested.

"It was," Dave said. "Did you try it?"

Cal shook his head. "Everyone else got a taste but me."

This time, Frida didn't laugh.

"How come you didn't get any?" she asked. "What were you doing?"

He told her about helping Rachel in the garden. "She told me the Vote is public," he said. "Did you know that?"

Frida seemed to think about this. "I guess so. I assumed it would be."

"Really?" Cal raised an eyebrow at Sailor. "Even at Plank, the controversial topics were voted by secret ballot."

Dave looked stunned, and Sailor gave him an appeasing look.

"We wouldn't know," Sailor said. "During our tenure, there was never a controversy we had a say in." He paused. "The school closing was never up for debate."

"I didn't know you guys went to Plank," Frida said. She

245

turned to Cal, as if to say, *Why didn't you tell me?* She didn't look angry, just surprised.

"You're such a fucking big mouth," Dave said to Sailor, who grinned.

"Keep thinking that, my friend," he said.

"Is it a secret?" Cal asked.

Sailor frowned. "The recruiter said no one would care where we came from. And that's turned out to be sort of true."

"Until now," Frida said, and reached out to push a lock of hair out of Cal's eyes.

Cal turned to Dave. "So if a bunch of you are Plankers, why not do things the Plank way and allow everyone to cast their decision privately?"

"Why?" Dave asked, eyebrows raised. "Are you assuming it's a controversial topic?"

"Yeah," Frida said, turning to Cal. "You think too highly of us, babe."

When they were alone on their walk back to the Hotel, far from anyone who might hear them, Cal told her to keep her pregnancy a secret. "Micah asked us to," he said.

He was surprised that Frida didn't protest, though he didn't say so. Instead, he began to tell her about his trip to the tree house. He waited for her to say that she'd been there, too, but she didn't. She didn't speak at all. It seemed so easy for her, to not tell him things.

He asked her if she felt different, now that she was pregnant. She just shook her head.

"Peter and Micah are looking for confirmation, I guess."

"So you want me to start barfing?"

He shook his head, and then nodded. She laughed, and relief moved like sunlight across his body. "You seem really happy today," he said finally. "Just now, when I saw you there, with Sailor."

"Anika knew Jane," she said.

"What?"

"Sandy had Jane on the Land."

"Are you sure?"

Frida nodded. "It's happened before, Cal. There's been a baby here."

She took Cal's hand and squeezed it three times, as if she were sending him a message.

"If it happened once," she said, "that means it can happen again, don't you think? Maybe that's what Micah and Peter were getting at. We just have to wait until we're fully accepted here."

They had almost reached the Hotel, where people were milling about. On the porch, a man was strumming a guitar with only two strings; Cal had learned his name yesterday but had already forgotten it.

"Frida," Cal whispered. "Be careful."

"Of what? Smolin, with his ballads?" She nodded at the man with the guitar.

Cal couldn't believe Frida was being so blind, but he didn't want to worry her or crush her hope. It was probably keeping her spirits up. He couldn't say what he wanted to say, which was that she might be wrong. Even if Anika was telling the truth, it didn't necessarily bode well for him and Frida. Sandy Miller might have had Jane on the Land, but Jane wasn't raised here. And what about Garrett? The Millers had left this place: that was the point. Now Cal and Frida needed to find out whether they had done so by choice.

"Don't get ahead of yourself," he said.

"Yes, sir," she said. After a moment she added, "Are you really joining Micah in the mornings now?"

"How does everyone know? This is why we can't tell anyone you're pregnant."

"It's true, then?"

He nodded. "I'll be on the inside, Frida."

"Try to hide your boner."

He ignored her. "I'll find out what happened to the Millers."

"Sure you will," she said, and raised an eyebrow.

They were almost to the porch, and Frida was waving at one person and then another, like a beauty queen on a parade float.

"You must be happy about the Plank contingent," she said.

"It kind of weirds me out, actually. What did they think they were coming out here to do?"

15

The cold weather had snuck up on her. Frost lay on the field every morning, and one night, hard rain pinged off the Hotel roof. The next day, the construction team had nailed boards across all the glassless windows on the Land. Now the Hotel was dank and fortresslike, Frida and Cal's room simultaneously stuffy and cozy, especially when they were falling asleep. "At least it's not freezing in here," Cal said.

Frida preferred the Hotel kitchen to anywhere else. Not only was it the warmest place on the Land, but she could also still look out the windows and watch the sky turn lighter and lighter as she worked. Once the sun rose, they had to stop baking and start Morning Labor, but she didn't mind. She was just happy to be able to walk down the Hotel stairs in the morning before anyone else did, pondering the tasks ahead and wondering if what she made would taste as good as what she'd served the day before. She had a reputation to uphold. After the first morning with the clafoutis, Anika had given in and allowed her to bake bread.

"I guess we need something to soak up all the soup," Frida had joked.

Anika didn't laugh. "We need bread for sustenance. Desserts are frivolous, but they help every once in a while to keep up morale."

Anika always had the oven lit by the time Frida met her in the kitchen. She'd be standing by it for warmth, and when Frida entered the room, Anika would lift one hand in greeting before bringing it back to the flames. More than once Frida had expected to see Anika plunge her whole arm into the oven with barely a wince; she seemed indomitable like that. Or just crazy.

Anika could be a little scary, but beneath her swagger was a softness. The more time Frida spent with her, the more it seemed that Anika longed to reveal this side of herself, exchanging history for history, secret for secret. She wanted to share things like old friends did, or maybe like a mother would, carrying her newborn through the house, naming all the objects around them. The lesson being: *This is how the world works. This is how we make order.*

It didn't take long for Frida to understand that Anika was a fine baker, probably a great one, and that she didn't need any help from Frida. Anika kept inviting Frida back to the kitchen, not for assistance, but because she wanted her there. She was after information, and Frida had it. For the first time, Frida was valuable.

They started out small. Anika asked, "What was it that Sandy named her second child again?" and Frida said, "Garrett."

"A boy." Anika paused. "How old?"

"He's four. Was. He was four when he died."

Anika nodded.

She waited for Anika to ask another question, and when Anika didn't, she realized it was her turn to ask something. It was that easy.

"How old was Jane when they left?"

"We'd celebrated her third birthday a few months before. I made her a belt I sewed from an old dress. It was purple, and adjustable because she was growing so quickly." She smiled.

"Everyone gave her presents, and we sang all the songs she loved."

Frida didn't respond immediately, and when Anika looked away, Frida felt the delicate connection between them tremble, threaten to snap. Frida realized she should have pretended to have seen Jane wearing the belt, but now it was too late to lie; Anika wouldn't fall for it.

During that evening's Church meeting, Micah told everyone that Cal was helping him in the mornings, as if everyone didn't already know. He also announced that he was pushing back the Vote until all the winter preparations were finished and August had returned from his latest trade rounds. August seemed to have left unexpectedly, but Frida wasn't sure how often he usually came and went.

"This will give you all more time to consider the decision," Micah said, and from the last row of pews someone yelled, "More time to eat that killer bread!" As far as Frida could tell, everyone laughed, even her brother. Anika grinned at her from across the aisle.

After the meeting, when she and Cal were lying in bed, Cal assured her that no one had complained about the postponement. "Everyone wants August here—his opinion matters."

"Where did he go this time?" she asked.

Cal said he didn't know the details.

Was that true? He still hadn't told her what happened during the meetings.

"You're so CIA," Frida teased. Let him be sly, she thought. He had no idea what she had planned; he had overlooked Anika, and Frida's mornings with her. Men were stupid to forget what good sleuths women could be.

The next morning, Anika brought a bag of coconut from the root cellar, along with the now-familiar baking crate. "This must have fallen behind the shelf," she said, and handed it to Frida. It was a plastic bag knotted closed, as if from a bulk bin, the flakes lab-coat white. Smelled like Thai soup, or like a high school girl's shampoo.

"I love coconut cake," Anika said, and took the bag from Frida. She shook it as if it were a snow globe. "There's a whole box of these bags downstairs. I totally forgot. August got them last time."

"Last time what?"

"On his last trip, months ago. To Pines."

Pines. Anika didn't even stumble over the word.

Frida didn't know much about Pines, except that it was one of the earlier Communities to be established, not long after Bronxville, Scottsdale, Amazon, and Walmart. It was the first to be named not for its original city or neighborhood, nor after the corporation that had put up the money to build its hospitals and schools, its borders and security teams. Its name was meant to summon images of nature and greenery. "And also stability," Toni had told her once. Pines was one of the smallest Communities, but it had a decent amount of money. Or it used to.

"I see," Frida said to Anika. She wondered if this was what Cal had been learning in the meetings.

With rounded cheeks, Anika blew the air out of her lungs. "We give August a list every couple of months, and he returns a few weeks later with everything we've asked for. Or almost everything, at least. It's been like this ever since Micah got here, though how he persuaded Pines to work with us, I have no idea. I wish I knew. Actually, no, I don't wish that. I don't want to know anything."

"Ignorance is bliss?" Frida asked.

"Something like that."

So that was how it worked. August went into a Community and returned bearing gifts. Was it like driving a car or sending an email, not having the least interest in how the science worked? Might as well be magic, because even if someone explained it to you, it still wouldn't make sense. Or was there another reason Anika preferred to be kept in the dark? Maybe it was dangerous to know how the Land worked.

Micah had once hated the Communities, and now he was trading with them. Frida wondered if Cal had pointed this out in the meetings.

Without speaking, Frida poured the bag of coconut into a bowl Anika had handed her, and swirled her fingers through the flakes. She'd never cared much for the taste, but she loved how it looked: as if a cake had grown fur. She imagined August buying the coconut from a supermarket in Pines. Did he use money? If so, where did he get it from? Why did they let him in? Were there even markets there?

Frida tried to remember what Toni might have told her, but she came up empty. Everyone on the Land had to know how August procured the supplies, but only a few would understand the process intimately. Cal might have learned about Pines days ago and kept it all from Frida, just like he'd hoarded Bo's story about the Spikes. There was no telling what he might keep from his little wife.

Frida had always been fascinated by the Communities, the secret life behind their walls, their riches and beauty all conjecture. In the first couple of years after they opened, Frida had conjectured a lot.

L.A. was a festering wound, but just a few miles away men and women slept peacefully on canopied beds in large rooms in large houses. At eighteen, Frida thought canopy beds were so glamorous. A few years later, when Toni started telling her about the Communities she'd researched, Frida had eaten up every detail: there were bikes everywhere, and helmets were required; residents had to pass a rigorous physical exam to gain entry; each child was sent a toy on their birthday. In a Community, someone flipped a switch, and a light turned on.

"You really don't want to know what goes on in Pines?" Frida asked.

"No, I don't." Anika raised her eyebrow. "Curiosity leads to trouble. You'll learn one of these days. What made you curious about the Land? Why did you come here?"

Frida had already told Anika about how she'd met Sandy and about her first visit to the Millers' house. Anika knew that Frida and Cal had been living there when they decided to come to the Land.

"Was it August?" Anika said. "He can be charming."

Frida laughed. "We didn't know he was here. We knew nothing about you guys." She described what Bo had told Cal: his story of him and Sandy visiting the Spikes, how they had turned back in fear.

Anika seemed confused. "So Bo acted like he didn't know us? He *lied?*"

"He didn't tell me, he told Cal. And Cal kept it from me for months and months."

"Men are asses," Anika said. "Stubborn."

Frida laughed again and smoothed the side of the last cake with a butter knife. They'd made five so that everyone on the Land could have a piece.

"He claimed he was just trying to protect me." As she spoke, an anger bit into her so deep she couldn't say anything else. Cal would withhold the world from her in the name of safety.

"Oh, I don't doubt it," Anika replied. "But that's always their reason. It was probably Bo's, too, because he was never a liar when I knew him."

Something in Anika's voice made Frida nervous. "What happened back then?" she asked. It was her turn for information, but she knew this was too broad a question, the parameters for answering recklessly wide.

Frida waited. 10-9-8-7-6 "Jane had friends here," Anika said. "She did?"

Jane had friends here. Jane had friends here. As she put the cakes in the oven, the words knocked deliriously around her brain like the lyrics to a pop song. Anika meant other kids, didn't she? She had to. Or was it simply code for having allies? Did someone not want the Millers, and Jane in particular, to leave? Anika could have merely been talking about herself. She had obviously adored that little girl.

Frida's stomach seized. It was as if she'd been struck with motion sickness, like she'd been reading in a car—she still remembered that feeling. She didn't take another step, telling herself that if she remained perfectly still, she wouldn't be sick.

She put her hand over her mouth as if to stop whatever might happen next. She must have looked green because Anika was right at her back. Frida vomited as she stepped into the small hallway off the kitchen.

"Sorry" was all she could think to say.

As soon as Frida had wiped her mouth with the back of her arm, Anika put her hand on Frida's forehead, then the back of her neck, asking if she felt hot, or cold, or a combination of both.

"I'm fine," she said. "It must have been the stuffy air in here."

If Anika was disgusted by what had just happened, she didn't show it. Frida was grateful.

Anika asked, "Does this happen a lot?"

Micah wanted her to keep her pregnancy a secret, but maybe the baby didn't want to cooperate. Thank goodness she wasn't showing yet.

Frida realized that Anika was judging her body: its strength, its health, its tendency to collapse into illness. It was how August used to treat her and Cal whenever he came to trade.

"Sometimes I get overheated," Frida said. "That's all."

"You should rest," Anika said. She nodded to a door across hall. "My room is right there. You can lie down. I'll clean this up and get you some water."

Frida could have hugged Anika right then, not only because she was being so nice but also because she was letting her inside her room, as if they were actually friends.

"Thank you," she said instead.

Alone on Anika's bed, Frida waited for her stomach to mutiny again. But the nausea had passed as quickly as it arrived. What a capricious little fetus.

So you're in there, Frida thought. If a baby could absorb nutrients from its mother's bloodstream, then it must be able to intuit her thoughts, too. *What number am I thinking of?*

She told herself, *Five,* and she imagined the baby holding up its paddle hand.

"I never doubted it," she said aloud, her voice lilting into song. If she wasn't careful, she'd soon be gaga-gooing to the tiny thing, whose heart couldn't be larger than a freckle.

She remembered Hilda saying that forty weeks was ample time to fall in love with a person you hadn't officially met yet, though when Micah was born, Dada said Hilda had been so tired and overwhelmed she'd beg him to take both kids off her hands. "Just for ten minutes, please," she'd say. "They're killing me."

"I hope there's just one of you in there," Frida whispered now.

She sat up and took in her surroundings. It was very similar to her own room, one of the few she had seen on the Land with a door. Many of the residences on the Land were wide open, without even a curtain to provide privacy. Fatima had given that up for them, which now struck Frida as extremely generous. But even their door didn't have a knob, and here was Anika's, with a knob made of metal, the kind Frida imagined in old Victorian mansions. About half a foot up from the knob was a modern-day lock, just like the one Cal and

Frida had had on their door in L.A.

How had Anika snagged that?

She looked around for anything else unusual, but there was just a bed and a child's step stool with a candle atop it. The single window was covered with a piece of sheer cloth; perhaps Anika would board it up herself when the cold became truly unbearable. Or, more likely, she'd suffer through the winter nights, teeth chattering. The closet did not have a door, but it didn't matter because the only thing in it was a pile of clothes, including the overalls Anika had been wearing a few days before.

Frida realized the nausea had distracted her from the best part of Anika's room—it was right underneath her. Unlike the straw monstrosity she and Cal slept on, Anika had a twin-sized mattress, practically new, maybe twenty years old. How had she gotten dibs on it? Anika must be favored. Not Micah-level special, but special nonetheless. The headboard was modern, too, cheap that way, made of a light, hollow metal, probably from Ikea, which had closed when Frida was twelve. "They took their meatballs and went back to Sweden," Hilda had said wistfully.

Frida turned onto her back and looked up at the ceiling. It was badly cracked, in worse shape than she expected, and she thought maybe she'd tell Cal about it. Maybe he could get

Micah to bring in the construction team. She realized how protective she'd become of Anika. She really cared about her.

Above her was a drawing. It had been made on a piece of fabric, cotton most likely, or maybe muslin, though Frida couldn't be sure, torn into a square and stuck to the ceiling with sewing needles. Only a soggy, sagging building would be weak enough to pierce with such flimsy things, Frida thought as she took in the drawing itself. It looked like charcoal, but more likely it was ash. There were two stick figures.

Jane had friends here. Jane had friends here.

Frida stood on the bed in order to get a better look. It looked like an adult and a child—both female, with triangles for skirts. To the left of them was a tree and, above them, a smiling sun. A few birds flew across the page, depicted as sure-handed Ms— had every child since the dawn of time learned to draw flying birds this way? Next to the figures was a tiny oval shape with eyes. Was that an animal? Or a baby?

Frida placed her hand on her stomach, finding her breath. She wanted to yank the drawing from the ceiling, find out more, a name maybe, but she knew she couldn't.

Was this a drawing of a mother and her daughter? Did she belong to Anika? Who but a mother would keep something like this?

Frida would ask her. That's what Anika wanted; she must. Tomorrow, as they baked, Frida would find out the truth.

There was a knock on the door, and Frida knew who it was before he stepped inside. Cal looked so clean compared with last week, when he would return from Morning Labor covered in dust and sweat. Now he wore the faded button-down jean shirt he had always loved. Holes in both elbows, but at least tucked in.

"You okay?" he asked, but not until he'd shut the door firmly behind him. "Anika came to get me."

"I've got those symptoms you asked for."

"I hope Anika doesn't put two and two together." He pulled the shirt from his waist, as if home from a long day at the office, and sat on the edge of the bed. "Just in case, Micah's telling people you used to barf a lot when you were a kid." He paused. "It's really happening, isn't it?"

She nodded. "I think so."

"No one can find out. Not yet."

Frida pushed herself up to sitting. "What happens if they do?"

Cal shrugged. "I don't know. That's why we should wait. I'm working on Micah."

"What do you mean?"

"To be honest, I'm not sure. Just trying to subtly persuade him, I guess."

Frida didn't want to laugh, but she couldn't help it. "My brother can't be persuaded."

"That's what I told Peter."

"You two talk about my brother?"

Cal nodded.

"Without him?"

Cal nodded again. "When we have the opportunity. Peter thinks Micah's a great leader, but that he needs to be kept in check. His ego, and all that."

"And you think you can change him?"

Cal shrugged. "I guess Peter got him to loosen up. He once stole Micah's clothes while Micah was taking a shower, then put them on one of the goats. I guess that finally broke your brother. He was buck naked and couldn't stop laughing. Maybe that convinced Micah he could think of this place as family."

Family. Frida put a hand on Cal's cheek.

"Tell me what you did in the meeting today," she said.

He smiled, but it looked more like a wince. "You know I can't."

Frida imagined Micah telling Cal about Pines. Her brother would describe in detail how August got inside and what he did once he was there. They were big secrets, she imagined, and they would have to be whispered. Frida couldn't summon Micah's words, though. It was like trying to continue a dream after she'd already woken, and she hated that she didn't know what would happen next. She wanted to know what Cal knew. Didn't she?

"You can tell me," she said. "I'm your wife."

He shook his head. "I want to, baby, you know that. But if they find out I did, I would lose the access I'm gaining. You have to understand."

"I see."

If he wasn't going to divulge, well, then, neither was she.

"I can't," he said. His words were sweet, but he wasn't even looking at her. "I crossed my heart and hoped to die."

"Stick a needle in your eye?"

"Is that really what the saying is? Jesus."

He was already standing, as if eager to be away from her.

———

Frida rested all day, and the next morning, when she got up to bake, the Hotel was still dark. She felt fine, thank goodness. Dawn was a ways off. Would morning sickness coincide with the rising of the sun?

She read the hallway walls with her hands, tiptoeing to the staircase, and wondered what she'd say once she reached the kitchen. She wanted to ask who had drawn that picture. She wanted to ask if Anika had been a mother. Frida wished she could tell Anika that she herself would be one in just a few months. Now, at least, she was certain of the pregnancy. Last night at dinner she'd refused the kale dish, the sight of greens making her queasy, and she'd fallen into bed soon after, as exhausted as she'd ever been. Her body was in this child's clutches, and he wasn't—*she* wasn't?—letting go.

If only she could tell Anika; Anika would understand. But no, Frida had promised.

They would make plain wheat bread today. No frills. Frida had decided last night. She and Anika had been too decadent lately, acting as if their reserves of chocolate, of coconut, of vanilla extract, were endless. If August went on a trip to Pines and returned empty-handed, she and Anika would be blamed for raiding the root cellar. Or she would be; Anika, special and feared, was probably above reproach. Yet another fancy dessert might imply irresponsibility to the rest of the Land, and that needed to be avoided.

The kitchen was dark when she reached it, the oven unlit. Where was Anika? Frida's heart hiccuped. She imagined the baby flipping inside her like a quarter, heads to tails.

"Hello?" she called out.

She hadn't told anyone about the drawing she'd seen in Anika's room, but when she'd left it, she hadn't closed the door, and someone might have gone in there uninvited. The drawing had to be a secret; why else hang it on the ceiling? What if someone had found out that Anika had told Frida about Pines? As an outsider, perhaps Frida wasn't supposed to know. Anika could be in trouble. Come to think of it, Frida hadn't seen her at dinner. *Don't panic,* Frida told herself. *Not yet.*

She wanted to run back upstairs and wake Cal. She would tell him everything in one long breathless rush, the same way he'd confessed Bo's story. That morning, weeks before, he'd led her back inside, and in the center of the very house the Millers had died in, he told her the truth. All of it. She wondered if that would ever happen again.

Now there were other people to consider.

She waited in the kitchen, her body alert and taut as a predator's. She could just make out the outline of a candle on the table and the box of matches next to it. Well. At least she could solve the first problem. Frida was striking a match against the strip of carbon when she heard the door open.

Anika walked in with a scarf around her neck and a glow stick in her hand. Frida had always imagined Anika walking through the Hotel in the dark, sniffing her way to her destination like a wolf seeking its dinner.

"You're late," Frida said. Wasn't that what Anika had said to her, that first morning?

"Sorry. I actually slept last night."

"You did? How long has it been?"

Instead of answering, Anika set to work lighting the other candles and getting the oven going. She unwound her scarf before the room was warm: a tiny form of penance.

"Did you dream?" Frida asked.

Anika shook her head. "Comatose."

"Why don't you ask August to get you some sleeping pills?"

"That stuff scares me. Anyway, I don't like pharmaceuticals."

"What about birth control?"

Anika smirked. "You're not wasting any time this morning."

"How can I, after what I saw in your room? That child's drawing."

Anika stuck another branch into the oven.

"When I was about six years old, my dog, a golden retriever mix, jumped onto the counter to get at a near-empty bag of

Cheetos. Remember those? They were chips, sort of. Orange and powdery."

Frida nodded.

"Well, no one was home, and Bongo got his face caught in the bag of Cheetos, and he couldn't get out. He suffocated in there."

"That's awful," Frida said. She meant it, but she couldn't help but smile. "And absurd."

"Curiosity kills, Frida."

"So does gluttony, apparently."

Anika shook her head. "You're not understanding me."

"Are you telling me there aren't any answers here? No Cheetos?"

Anika held up a branch as if she were considering displaying it on a mantel. "It's been so long since everything happened, I wonder if it means anything anymore."

"It means something to you. I can tell."

Anika threw the branch into the fire, which was strong and hot by now. "Let's get started," she said. "We're running late."

Frida told her she wanted to make bread, nothing fancy, and Anika didn't offer her opinion, as she usually did.

The bread didn't take much time to prepare; it was so easy, it was hard to ignore the truth: that their morning baking was just an excuse to share stories. A ruse.

"Okay," Anika said finally. "I do want to talk about it."

"Then go ahead," Frida replied. "I'm a pretty good listener."

"But you've kept the most important thing from me."

"I have?" Frida said. She had the large bowl of dough in her hand. It was ready to rise, and she held on to it tightly. Anika knew she was pregnant. She must have guessed from Frida's vomiting. Or maybe she had heard her and Cal whispering in her bedroom. Maybe everyone knew.

"The Millers," Anika said.

Frida exhaled. Her friend was just as oblivious as ever. "What about them?"

Anika brought her voice to a whisper. "As soon as he heard, August told me they were dead. He would never keep something like that from me." She paused. "But why? Why did they do it?"

"Anika, I have no idea."

"Bullshit. Tell me why they killed themselves."

Frida put down the bowl. In an hour the dough would be ready to knead. After only a few days of working in the kitchen, Frida's hands and arms were baker-strong again, able to make loaf after loaf of bread for the Land. The body never forgets.

"I swear I don't know, Anika. It's haunted me for months."

"When August told me the news," Anika said, "I nearly fell down. He had to hold me up."

"Cal buried them," she said. "It was horrible."

"You know what's horrible? That I thought you'd have something useful to tell me."

"I'm sorry."

"All this time, I hoped you'd be able to shed some light on what happened, give me some solace. Help me understand. We allowed Micah into this place so that he'd protect us, and he let two of our founding members die."

"They weren't here anymore to be protected," Frida said. "Why did they leave?"

Anika didn't reply.

"What difference would it make if I knew anything?" Frida asked. She grabbed Anika's wrist. It felt wrong, like putting her hands on a stranger.

"This isn't a fair exchange," Anika said.

"No, it isn't." Frida didn't say that it reminded her of

marriage, which was never fair, but at least it always changed. You gave and gave and gave, and then, eventually, you found yourself taking. Which was the better side to be on?

Anika hadn't moved her wrist from Frida's grasp, and Frida took this as a good sign. "Tell me about the drawing," she said, her tone almost imperious. She sounded like Anika herself, and Anika obeyed.

"I had a boy," she said. "That funny little egg shape in the picture? With the eyes? That's my baby. Jane drew it for me— it's of the three of us. I was like an aunt to that girl."

She had her son late, as far as those things went. Forty-four. When she missed her period, she initially thought maybe she was going menopausal. But in June, on the longest day of the year, "literally and figuratively," Anika said with pride in her eyes, she gave birth. The labor took forty-two hours; he was born in the barn, like an animal. They named him Ogden. "Not after the poet," Anika said, but Frida had no idea who she was talking about. Cal would, but she didn't think she could tell him this story.

"Was he the first child to be born here?" Frida asked.

"We weren't the only ones, Frida."

The Land used to have families. It wasn't teeming with kids, but there were about fifteen or so when Micah arrived.

"There were children here when the Pirates attacked?" Frida said. "How could you leave something like that out?"

"I'm sorry," Anika replied. "Sometimes . . . it's too much."

"Were they hurt by the Pirates?"

"No, not physically. But they were just as afraid as we were, if not more so."

Frida kept her voice gentle. "Tell me about the children. You have to."

All but three had been born here; the oldest, Melissa, was

twelve and had come with her parents. The girl remembered her life before: the crime, the hunger, how the city had been promised paved roads, schoolbooks, medicine, but they never arrived. She was five when her parents decided to follow some old friends out here. "They'd come from Merced, so you can imagine."

In the bowl, the dough was rising. Frida imagined it inflating like lungs.

"A few babies died," Anika said. "And we lost one woman to childbirth. That was in the beginning, before we had a handle on agriculture, when our nutrition was poor. I mean, our nutrition was never great, but in the beginning, it was the worst. If a pregnant woman doesn't get enough protein, she's at risk of hemorrhaging when she gives birth." Anika sighed. "It's harsh out here, for anyone. But for a kid? Melissa had been sick with the flu for days when Micah showed up."

Her brother. This was where the story changed.

"Melissa," Anika said, "she—"

"What happened when my brother got here?"

"Don't interrupt me."

"I'm sorry—"

"He arrived in the carriage one evening, right after dark. With August and a handful of other men: Burke, Sailor, Dave. None of the others yet—they would come later. Three weeks earlier, before sunrise, a group of Pirates had come."

The Pirates had raided their reserves of food. They'd tied a red flag to the steeple as a warning, but no one saw it until afterward. "Almost all of us were still asleep when they rode up. I was awake, just about to head with Ogden to the river to do laundry. I had him strapped to my chest, and it was cold, so my hands were wrapped in scarves, to keep them warm. I don't know why I remember that. I never do that now, haven't since that morning.

266

"As soon as I heard their horses' hooves pounding against the ground, I screamed until my voice went hoarse, my hands over Ogden's ears. He started crying but miraculously stopped a moment later, as if he knew not to draw attention to himself.

"This time, the man with the scaly arms, the long-haired one, wasn't there. A new Pirate was in charge. He was young, he couldn't have been more than eighteen, and very tall. I remember thinking he was taller than any of the men on the Land. It looked like he'd tried to grow a mustache, but it didn't take, it was just a bunch of dark scraggly threads. Not that it mattered, he was in charge, all the others followed his orders. There was something horrible and unfeeling about him, and I knew right away that things were going to be worse for us now that he was running the show. He trained his gun on me as his men pulled the others out of the Hotel and lined them up. One of the women tried to run, and she got clocked in the face."

Frida shuddered. "What did they take this time?"

"What they took doesn't matter," Anika said. "It's who."

"Another woman?"

Anika shook her head. "The Pirate, the one who was now in charge, he got off his horse and walked down the line of us. He was wearing old black basketball sneakers, the toes mended with duct tape. I kept my eyes on the shoes as they moved from person to person. Even the kids knew to be still. It was quiet except for the wind, which was blowing pretty hard. Some door was slapping open and closed."

Frida said nothing.

"Randy was eleven years old then, big for his age. He was the second-oldest kid and had been born in L.A. We sometimes joked that someday he and Melissa would get married."

Anika smiled, but for only a moment.

"When the Pirate got to him, he stopped. Randy's father was

267

one of the first men killed by the Pirates, and, I don't know, maybe that marked him. I remember Deborah, Randy's mother, was farther down the line. When the Pirate nodded at Randy and pulled him out of line, Deborah saw what was happening and screamed. The babies started crying—but not Ogden, though. I remember whispering to him that we'd be all right. But, really, I was so scared.

"When a few of our men tried to step in and stop what was happening, they were shoved to the ground. Deborah ran toward her son, but one of the other Pirates tackled her. She would've let them kill her, except her son got right on the Pirate's horse, as instructed. The boy didn't say anything. Maybe he already knew what it all meant, that he was being recruited into their terrible army. And maybe his silence was what kept the rest of us in line. Randy was looking down at the horse, as if he'd never touched a saddle. It was like he was being sacrificed for our safety, and he'd accepted it."

"He just let them take him?"

Anika didn't answer.

"Anika?"

"We told Deborah that Randy was a hero for giving himself up like that, for the good of all of us. What else could we have said?

"We'd been so terrorized, that when Micah and the others showed up three weeks later, we were afraid. I came running at them with the scythe, screaming once more. They could slice out my vocal cords if they wanted to, I didn't care. Micah held up his hands in surrender, and I was caught off guard. He had a pear in his hand. A peace offering."

A few hours later in the Church, August asked them why they hadn't fought off the Pirates. "We said we'd tried, but that we didn't have the manpower or any more weapons. Micah did, and he had twenty-five more men and women waiting nearby, ready to come to the Land to protect us. He had a solution."

The Pirates were using the Land, he said. "Unless there was a real threat against them, they would continue to steal from us until we didn't produce enough to make it worthwhile, and then we'd be unceremoniously murdered, every last one of us. I remember he'd looked at Deborah and said, 'But before that happens, they'll take the stronger boys. I don't want to say what they'll do to the girls when they get older.'"

Micah and his people were looking for a place to settle. "They needed a permanent territory, and we were it," Anika went on. "Back then, there had been a few Forms; they were built long before we came around, probably by some eccentric artist. Did you see them on your way in—they're smaller, but no less frightening."

Frida shook her head.

"Those original Forms gave Micah an idea. 'It's a natural border,' he told us. 'And you'll be protected.' Peter wanted to know more."

"Peter was here all along?"

"One of the first." Anika turned to Frida then and gave her a soft smile, as if she were a kind but firm boss about to lay off her least-productive employee.

"Why are you looking at me like that?" Frida said.

"I suppose it's time I tell you that Peter was Ogden's father."

A silly gasp escaped from Frida's mouth. "You and Peter?"

"It's long over," she said. "It ended even before I gave birth."

"So you guys came here together."

"No, he came from Portland with a few of the others. We were from all over, though. I'm from Oakland. A bunch of us communicated online years before, and then we finally decided to meet and make a go of it. Not that we knew what we were

269

doing. We came out here without any conception of what we'd need to survive." She shook her head, a rueful look on her face. "We barely had enough food. We lived in constant fear of Pirates. Melissa was very sick, and when Marie's milk dried up a few weeks earlier, her baby had nearly died of starvation until another woman got him to eat some sweet potato.

"I remember Peter's arguments well, because they were articulated so clearly, it was almost impossible to find fault with them. He said we'd all perish if something didn't change. We had to welcome Micah and his gang of settlers. They had skills; they would help us. He said that Micah's plan meant safety for everyone. Peter's confidence in Micah convinced everyone else."

Frida nodded.

"The Pirates only returned to the Land once more. We'd started security shifts, and Micah told us to let that little shit leave the red warning, pretend we hadn't seen him come and drop the bandanna in front of the Hotel in the middle of the night. Micah didn't want them to know anything had changed, and that was smart." She sighed. "Two days later, when the Pirates finally arrived, your brother stepped out of the house he'd claimed as his own, and without even a word he shot the young leader in the chest. The kid gasped, and he fell off the horse. I remember the way his long body hit the ground: sort of dripping off the animal. Randy was with them, on his own horse, his hair matted into knots, shirtless and in a pair of red gym shorts. He was sunburned, and he had what looked like cigarette burns all over his arms.

"Micah told the others to hand over Randy right away. I think the men were shocked. When Randy didn't move, August and Sailor came out from behind another building and shot his horse. Randy fell off. For a moment, he wasn't sure what to do. Micah said to stand up. As he spoke, some of our men emerged from other houses, all of them with guns. The Pirates who were left whistled once and turned around. I guess the new kid wasn't worth a fight. I bet one of them was ready to take the throne after the assassination.

Frida wasn't sure if this was surprising or predictable. Her brother was capable of shooting a man point-blank in the chest. She should have known. That's what Cal might say. The Group had inured him to violence, made it into a game. And these Pirates had done unspeakable acts to the Land, and taken one of their children. She remembered her brother as a kid: long eyelashes, arms thin and lanky, pen marks all over his hands from drawing on the butcher paper Dada had bought them. Micah had never been interested in guns and shoot-'em-up games. He'd been sweet.

"You should know, Frida. Micah beheaded that man."

"He *what?*"

"He called us all over and told us to watch. He pulled the body from the ground, and with his knife sliced his head off at the neck." She paused. "If that sounds like it was quick, it wasn't. The head didn't come off easily, and your brother had to saw across bone and tendons until it was removed. One of the babies didn't understand what was happening and was squealing with laughter. I remember Sandy covered Jane's eyes—but Sandy didn't keep herself from looking. None of us did. What Micah was doing was horrible, but that man had taken Randy. We *wanted* to see it.

"Deborah had grabbed for Randy as soon as she could, but when Micah started, Randy began to cry and pull away from her, as if the Pirate were his father. Micah ordered Randy to follow him out to a Form and made him hang up the head. Micah left it to rot into a skull. There were a few more skirmishes with Pirates after that, but never on the Land proper. By the time they stopped altogether, there were five heads hanging from the Forms."

Neither woman spoke. Frida searched for something harmless to ask.

"You said you built more Forms?"

"We did. We hadn't thought of them as security, as a gate to keep people out, until Micah suggested it. We also built the lookout Towers. We worked for months. Aside from using all the inessential stuff here on the Land to add to the Forms, August had begun leaving and coming back with discarded items for us to use. He was going to Pines—that's when we found out about that. I remember my hands then; they were shredded from all the manual labor. I was glad the kids missed that."

"You haven't told me what happened to them. Where is your son?"

Anika looked like she might throw up, and Frida glanced at the window to check the dawn's progress. The sun was about to rise. When Anika spoke, her voice was low and quiet. "It's almost dawn. Morning Labor is about to start."

"Just hurry and tell me now."

Anika shook her head and grabbed the bowl of dough. "You can't make me rush a story like that. Let's knead this before it's too late."

16

A t the morning meeting, they convened in a circle on the Church's stage. It was so cold, they covered the icy metal chairs with blankets before sitting down. August had returned the night before; now he held a clipboard stuffed with yellowed loose-leaf paper and a pencil that he'd sharpened first with a knife and then, when that wasn't quite sufficient, his teeth. Cal had to admit, the guy did look pretty tough gnawing at the lead. As usual, they spent half an hour reviewing Labor assignments, all of them outlined in August's notes, and then Sailor told them about his meetings with the team leaders. Cal had been deemed a strong critical thinker by the construction team leader, though he was "unnervingly quiet." Sailor raised an eyebrow at Cal as he said this. "And everyone in the kitchen loves your wife," he added.

"What's going on with the baking?" Micah asked.

"Please don't stop them," Sailor said quickly. "That coconut cake she made, and her sourdough? I mean—wow."

"It didn't occur to me that I *should* stop them," Micah said. "As long as Frida wants to keep doing it, that is."

"I'm sure Anika has a whole baking regimen in place—kneading, baking, calisthenics," August said. "Poor Frida, I wouldn't want to work for that woman."

Cal wanted to say that his wife enjoyed her baking sessions

273

with Anika, but he held himself back. Frida hadn't had a friend in so long, not since Sandy Miller, and it was obvious that her time in the kitchen had helped her. To Cal, Anika seemed stiff and humorless, but Frida could draw out the fun in anyone. Hell, she'd done it with him.

"Makes me nervous," Peter said, and Micah looked up at him quickly. Cal thought he detected a slight shake of Micah's head, or maybe in his eyes, a speechless *no*. He wasn't sure what they were worried about: that Frida would tell Anika about her pregnancy?

August changed the subject so deftly, Cal hardly caught that he was doing so. In moments they were onto other mundane matters: who wasn't cleaning up after themselves in the Bath; what still needed to be done for winterizing; if there was enough meat on Snorts, one of the pigs, to butcher him.

When they got to questions of agriculture, Cal leaned forward. These past few days, he'd found himself loving this part of the meeting. It made him think of his job back in L.A., working with the volunteers to make sure the crops they'd planted were thriving.

"We want to reorganize the garden next spring," Peter said, "but we have no real plan of action."

Cal realized that everyone was looking at him.

"It's a mess," Dave said, and Sailor groaned. "If we lose another crop of lettuce, I'm going to—"

"Plus, the seeds," August said.

"Sailor," Micah said. "Tell Cal what you've got."

He had heirloom seeds, Sailor explained, from his uncle in Charleston, South Carolina.

"Stuff you haven't ever seen before, stuff that hasn't been grown commercially for three, four hundred years," he said. "He gave them to me when I left for Plank. I hoarded them when

I first got there, don't know why. But I brought them here. We should use them."

Cal couldn't help but feel giddy. He'd read about seeds like these.

"I'd be happy to take a look."

The men, even Micah, beamed. That was it then: they needed a farmer for their Village People. This was why Cal had been invited into the circle of power.

They'd moved onto plumbing. There was a question of making the work mandatory for everyone, including themselves. "If we do it without complaint," Peter said, "it'll set an example."

All the men begrudgingly agreed.

"The truth is," August said, "the job can't be voluntary anymore. Nobody wants to do it."

"Latrine digging isn't that bad, especially compared to maintaining the outhouse," Dave said, and Micah held up a hand, wincing.

August turned to Cal. "This is glamorous, isn't it?"

"Certainly is," Cal replied, and August laughed.

Cal was relieved when the meeting ended. Even the discussion of security, plans to build three more Forms, and adding another man to the night shifts proved a snore. Part of him was glad the meeting had been so boring; he wouldn't be compelled to talk about its tedium with Frida. When he'd finally learned about Pines two days earlier, about how August traded with them, Cal had been glad for the keep-quiet rule. It kept him from repeating to Frida what she might not be able to hear. It had been a relief.

Before Cal left the meeting, Micah winked and said, "Fun, right?"

*

Micah had used that word, *fun,* so carelessly. The meetings weren't about fun. They were an opportunity: Cal had been invited to peer behind the curtain. At least Cal was learning how the Land worked. Frida didn't seem that interested anyway.

The truth was Cal was biding his time. He had plans to ask about the recruitment process. It seemed a harmless topic, outdated as far as he could tell, since the Land's philosophy was about containment, not expansion. He wanted to know who had lured Dave and Sailor here and why.

Toni had been the one to tell Micah and Cal about the Group. It was during her visit to Plank, after they'd gone to the street fair. Her cousin had eaten two funnel cakes and a fried candy bar, and as soon as they pulled onto campus he'd opened his door to throw up a murky soup that made Toni ask if she should call someone for help. To Cal's delight, the boy had shaken his head and stumbled away, leaving the two roommates alone with the girl every other Planker was dying to talk to. Or at least smell. Someone had whispered it was jasmine oil she wore; someone else had seen her shampoo bottle in the upstairs bathroom. "Peppermint bark," he'd reported, and the news was whispered from one ear to the next.

In other circumstances, Cal probably wouldn't have given Toni a second glance, but this was Plank. No female students. No Internet porn. No neighborhood girls to fantasize about. Not even a female professor since he'd started. Sometimes it felt like Cal's desire would eat him alive; he fell in love with trees, a certain horse, a washcloth in the farmhouse bath. Not that masturbation helped. That was just the body, turning inside out.

Toni was short and athletic, no breasts to speak of, her long brown hair tied into a ponytail at the top of her head. She seemed to like Cal, but only in the way you might like a boy you were babysitting: because he said funny things and went to bed on time. With Micah, she was different. Cal could tell by the way she laughed at Micah's jokes, and the way her voice went higher when she addressed him, as if she wanted to sound more feminine. She seemed to drink in everything he said.

The three of them were leaning against the car, staring at the stars. Toni had asked Micah to point out constellations, and the only reason Cal was still there was because Micah didn't know shit about stars and had said so.

"I hope my cousin won't get you in trouble for going to the fair," Toni said to him.

Micah nodded. "For that stunt? Definitely. We'll probably have to clean up goat poop for three weeks straight." He was lying, of course, but Toni didn't need to know that.

"Reminds me of some friends I have in L.A. In Echo Park. They're always pulling these crazy stunts," she said.

She and Micah commenced a conversation about neighborhoods and geography that, at the time, didn't make sense to Cal. He had looked back up at the stars, and the stars behind those stars, and then at the farmhouse. He thought he saw Plankers, perched at their windows, watching them.

"Anyway, my friends do this thing," Toni said, raising her voice a little. She wanted to bring Cal back into the conversation, he realized.

"They call themselves the Group," she had said.

"The Group?" Cal repeated.

Micah crossed his arms.

Her friends had created it, Toni explained; she'd met them while hitchhiking from Seattle, her hometown. "Or maybe they aren't the founders. It's all very nebulous," she said. "It's a performance group, but with a political edge. They do some amazing, thought-provoking stuff, and so far they've managed to get away with it. It's probably because people aren't sure who they are, but everyone loves their stunts."

Micah had uncrossed his arms by then. "What kind of stuff do they do?"

"It's better if you see for yourself. Get to a computer," she said, "and look them up."

She and Micah were standing closer now. Cal knew when to make himself scarce. He interrupted their conversation to thank Toni for taking him to the fair. "Have a nice night," he'd called as he walked away, but they weren't listening.

That night, Cal slept in the stable beneath a blanket that smelled like hay. He never thought he'd see Toni again, and he didn't think his roommate would, either. But after that, Micah found a way to get off campus and to a computer and learn about the Group, and he and Toni started a correspondence.

Cal preferred not to think about the rest of that semester. How Micah began reading Guy Debord and a slew of other French writers, then some anarchists Cal couldn't keep track of. There was even a small blue book written by an anonymous committee of writers and, predictably, the continual rereading of Marx. And then, who knew what else? Micah began asking others at dinner, "What do *you* believe in?" He was excited about returning to his hometown and seeking out the Group. "How will you make money?" Cal had asked once. Micah shrugged as if the question, and Cal's pragmatism, bored him.

After they'd both moved to L.A., Micah invited Cal to a meeting of the Group. Cal had told himself he wasn't interested in the Group's stunts. He wasn't, but he also didn't want to see Toni. Not that he carried a torch for her or anything; it was just that he didn't want to feel like a little boy around her: blushing when she spoke to him, feeling jealous when she paid attention to anyone else. Cal didn't know then that Toni and Micah would become

278

an item, and that she and Frida would become friends. Frida had no idea they'd met Toni at Plank; Micah and Toni had asked Cal to keep that secret, for reasons that Cal didn't think too much about. But every once in a while it was hard not to ask himself the uncomfortable question: if Toni had been interested in Cal back at Plank, would he have joined the Group instead of Micah?

Now, Cal couldn't stop imagining Toni marching onto the Land, unannounced. She had disappeared from L.A.; maybe she would reappear here, just as Micah had. Cal and Frida would have a chance to find out what had happened to her.

Perhaps she was the recruiter, and always had been.

He would ask Micah and the others, and they would tell him because he was one of them now. He'd just need to give it a little while. He reminded himself to remain silent, to listen, until the right moment presented itself.

———

The next morning, August held up a new secretarial prop, this time an old-school reporter's notepad, which fit like a Device in the palm of his hand. "It's time to start plans for the next journey. I'll probably go in a few weeks, so I need to get another list going. Of needs."

Cal had quickly learned that a *journey* meant a trip to Pines. Not to be confused with a *round,* which referred to August's regular survey of the areas around the Land. From what Cal had gathered, there were only a few settlers scattered in a hundred-mile radius. All were peaceful, and few interacted with one another. On a round, August's job was to make sure the settlers stayed put. They were not to go exploring. On a journey, August avoided these settlers; he had but one goal, and it was to get to the Community.

"They've asked us to report on the rate of disease," Micah said.

"What's the concern?" Peter asked.

"Probably isn't one," Micah replied. "They just want to feel smug that their mortality rate has leveled off. You know, that they're truly protected."

"I'm sure they're vigilant about possible viruses," Sailor ventured. No one responded.

"Let's ask them, straight up, why they need that information," August finally said to Micah. "They've been more forthcoming lately." Micah nodded. August held up the reporter's notebook and said, "And what do we need from them?"

Sailor leaned forward, his eyes rolled up to the ceiling, as if he were trying to recall the capital of some far-off country. "Housekeeping needs vinegar and another set of mops, if we can get them, and we're low on iodine tablets. Anika wants the baking crate replenished."

"Of course she does," Dave said, and snorted.

"She does have a lot of requests," August said. "Remember when we got her ChapStick with SPF? It was hard to get one tube, but once she had it, all the other women wanted one for themselves? We don't want another situation like that on our hands.

"This list will be tough, too. Last time I was there, it wasn't easy getting the iodine tablets. The demand is so high. They're paranoid about the water supply. Lord knows why, with their filtration system."

"Water at Pines tastes like a motherfucking swimming pool," Micah said.

Everyone laughed except Cal. He knew he shouldn't be shocked by the conversation, but this was the first time he'd witnessed the men make plans for a journey. They were being so casual about it. And why not? The facts seemed simple: August trekked to a Community, to Pines, to procure the Land's supplies. And to get those supplies? They provided information: the rate of disease, who might be a threat, who was out here. Could that be all? It seemed too easy.

August said they could revisit the list of needs the next day, and the men fell silent.

"Now that August's back," Micah said after a moment, "the Vote is upon us."

"Only one or two are uncertain," Sailor said.

"It should be unanimous," Peter said. "Even if it doesn't have to be."

Cal didn't want to know how Micah planned to convince people to vote a particular way. Something to do with denying requests for supplies from Pines, perhaps.

On his way out of the Church, Peter called Cal's name.

Cal hung back and let the others step outside ahead of them. Once the two were alone, Peter asked, "How's Frida?"

"Great. She really likes Anika, actually. I know you guys think she's sort of schoolmarmish, but she's been welcoming to Frida."

"I see," Peter said.

"What?"

"Nothing. Just let me know if anything changes." He ran a hand along his chin, which was peppered with stubble. "She's a tricky one," he said after a moment, and Cal didn't know if he meant Anika or Frida. "Any changes in Frida, physically speaking?"

"I think she's showing, but she says I'm nuts."

"You are. She isn't."

It made Cal feel strange, imagining Peter watching Frida, eyeing her body for that tell-tale rise above her belly button. It wasn't his job.

"You'd like a kid here," Cal said. It seemed obvious to him now.

Peter laughed, but it was too big, too horsey, to be convincing.

"Everyone would," Cal added.

"Not everyone."

"I've thought about what you told me before," Cal said. "How Micah can change."

Peter had started walking toward the open door. "It's not that he can or can't change," he said quietly. "It's more that he isn't the person he projects himself to be. Not exactly."

"No shit. He faked being a martyr."

Peter shook his head. "No, I mean, he isn't only serious, he isn't only tough." He smiled. "Just look at how he treats Frida. With her, he's a softie."

"Only with her."

"Maybe, but maybe not."

Cal thought they were done talking, but then Peter said, "You're acting differently around Frida." He spoke so earnestly, Cal couldn't be angry.

"I am?"

"You seem distracted, like you aren't paying attention to her. You don't want to make her feel vulnerable, not at a time like this. She needs to know you're there for her, so that she doesn't confide in anyone else."

"She knows."

"I don't mean to pry," Peter said. "We just want to make sure she's happy. That she's ready for the Vote." He seemed about to say something else but stayed silent.

Once they'd reached the doorway, Cal had to close his eyes against the sun. When he opened them again, Peter was already walking away.

*

Cal didn't want to admit it, but recently he'd become so consumed by the meetings, he'd stopped thinking much about Frida. He still thought about their baby, suspended impossibly within her like a galaxy of color in a glass marble, and about the Vote, what it might mean for them, their future. But that was their life, something shared, handed back and forth between them. Frida herself had drifted from his mind. He didn't think about whom she might be hanging out with when they weren't together, or how she was feeling, or even how she looked: someone had trimmed her hair and given her a pair of men's boots to wear, and he hadn't noticed either until she pointed them out. "Hello?" she'd said, tapping her toe, pretending to be upset by his absentmindedness. Or pretending to be pretending. He could see that she was a little hurt by his cluelessness. It had been days since he'd really looked at her.

Just the night before, she'd nuzzled into him in bed, and he'd said, "I've got my period," which was their shorthand for being too tired for sex. She'd laughed and given a fake whimper, and he'd wondered if there was something truthful to her little cry. He'd grabbed her hand then, so that she wouldn't feel rejected. He must have fallen asleep soon after, though, because he couldn't remember what had happened next.

Someone had to have let go of the other's hand first.

Peter was right; he was distracted, as stupid as it was. At first he was consumed by the connections between Plank and the Group, and he'd started to wonder if Toni had recruited Sailor and the others. Now, he was trying to understand the Land's connection to Pines. He didn't know what he'd tell Frida or if he'd tell her at all. He was still trying to make sense of it himself.

He thought about what Micah and the others had said about Anika. Frida probably had no idea that the men were concerned about the women's morning baking sessions. They didn't want

Frida to confide in Anika about the baby. Was that all? Whatever their suspicions, Frida would go on being unaware of them because Cal wouldn't say anything. He felt a prick of guilt at that, but nothing more. He didn't want to tell her. Why alarm her? Besides, if Frida was as cunning as she thought she was, she'd figure out their suspicions on her own. And if he told her, she'd probably just laugh. "Do they think we're getting freaky in there?" she might say, and brush off the men's concerns as silly.

On his way back to the Hotel, Dave and Sailor stopped him to ask if he wanted to join them on security after dinner. "It's the night shift," Sailor said. "So you better take a nap."

"Sounds good," he said, and smiled.

He tried not to seem too eager. He remembered his arrival here almost two weeks before, how he and Frida had rounded one Form and then another, his mouth so dry it felt like he'd swallowed a handful of pebbles. And Sailor, stepping forward with his fake-brave grimace, like someone's pest of a little brother, piggybacking on a game of cowboys and Indians.

If he and Frida were voted out and they had to leave the Land, how would they become an army of two again? Three, with the child. At first, Cal couldn't wait to get away from these people, from her brother. Now he found himself happy to awake on their sharp hay mattress, thoughts of the meetings, of the Land, filling his mind.

Micah, that sneaky bastard. He knew Cal would like it here. It *was* fun.

He went to find Frida in the kitchen. As soon as he saw her, standing at the worktable before a pile of chopped onions, he remembered what Peter had said about making sure she didn't feel vulnerable. She did look a little lonely, he thought, as she pushed the pile of glistening, weeping onions into a small hill.

He called out her name.

In front of everyone, he put his hands on her shoulders. With his thumbs he searched for the knots along her shoulder blades; there they were. He was a terrible masseuse—she'd always complained about it—but that didn't stop him from trying.

"Let's go to the Bath," he said.

Burke lifted both eyebrows and said, "Nice."

"Get your mind out of the gutter," Frida said to him, and Cal felt himself blush. She slipped from his embrace and turned around. With a hand to his cheek, she held her eyes on him for a moment longer than usual.

"Those onions are strong," he said, his eyes tearing.

"Aren't they?"

Cal tried to put his arm around her on the walk over, but she said she wanted to wash up first. "I smell too much like onions," she said. "You better stay away." He nodded, and they were silent the rest of the way.

Even once they were alone in the Bath, they didn't speak. Cal tried to figure out what to say.

"You okay?" she asked after a few minutes, as if she were waiting for him to unleash the darkest contents of his soul.

Cal told her what he was doing that night.

"Security?" she repeated. After a moment she said, "You know, only men are on that detail."

Frida was tweezing her eyebrows. As she spoke she looked at her reflection in a hand-mirror that had been taped to a bowling pin. Whereas in the kitchen she'd been keen to meet his gaze, now it was if she couldn't pull herself away from her own reflection.

"It makes sense," Cal replied, eyes on his fingernails. He loved the clippers on the Land; it was so much better than using his teeth to bite his nails down. The clippers were sharp and precise, with a little plastic reservoir that caught the cut nails.

"It does?" Frida said, finally looking up. "Cal. I weigh more than Sailor."

"It's about upper-body strength."

"To fire a gun?"

Her voice was louder now, and when Cal spoke again he was careful not to match her volume, even though he knew she hated how calm he could be. *Unflappable* was the word he'd use to describe himself, but she had once called him *robotic*.

"Would you even want to stay up all night to watch nothing happen?" he asked.

"Probably not," she said. "But why would my brother be interested in a male-run world? Hasn't history taught him anything?"

Cal caught her gaze in the mirror. "What's got you so riled up?" He put down the clippers and sat up straight. "Do the other women feel the way you do?"

Frida lifted the tweezers to her face. If she wasn't careful with those, her brows would be as curved as his fingernail clippings.

"You think I'd tell you what the women think?" she said. "You're a narc."

"And you like that about me," Cal said. "Admit it."

She didn't smile. "We're doing this a lot lately."

"Doing what?"

"Bantering. Bickering, really."

She was right. "I guess I hadn't noticed," he said.

"What *have* you noticed? Or wait, no, don't tell me. You *can't* tell me."

Cal looked away. All the concern he'd had for her in the kitchen, that she might be feeling alone, that she felt tense, seemed suddenly inconsequential next to his annoyance, his anger. She'd been the one treating the Land like sleepaway camp. She'd been the one to make him look like a moron in front of her brother.

"How many times do we have to go over this, Frida?" he asked.

"Go over what? We hardly talk."

"Do you want me to stop going to the morning meetings? Is that it?"

"Today when you grabbed my shoulders," she said, "I thought I might faint, it felt so strange. You never look at me. You hardly touch me."

"Stop being so dramatic," he said, and he stopped before he said anything meaner. He made his voice as even as he could. "I'm going to make things okay. You have to listen to me and promise me that you won't say anything about the pregnancy."

"I already promised you that."

"You have to trust me. If you don't trust me, none of this will work."

"Do we want it to work?" Frida said. She was still seated at the mirror, and she looked up at him, tweezers poised. "Every day our child becomes . . . I don't know . . . more and more *itself*. More part of us, Cal." She paused. "At first, I didn't even care if it lived or died, as long as we got to stay here."

"Don't say that," he said. If she wanted him angry, if she wanted his voice to fill this tiny space, well, he'd give her that.

"It's not how I feel now. Now I'm more and more nervous about the Land."

"Baby," he said.

She let go of the tweezers, and they clinked against the counter with a little pitiful sound. She said his name, so quietly he thought he might have imagined it.

"What is it?" he asked.

"Don't you know what went on here?" she whispered.

Micah and the other men didn't spend much time talking about the past: their meetings were obsessed with day-to-day details, with securing communication with Pines and making sure the settlers around them behaved, stayed put, didn't try to conquer the area or head for Pines. Cal had never thought to ask about what had gone on before. Before what?

"What did you find out?" Cal asked softly. "You can tell me."

She shook her head. "Can I?"

It was the worst thing she could say, and they both knew it. She didn't trust him.

"Micah doesn't seem concerned with whatever used to go on here," he said. "That's in the past."

"It is? You sound like someone else, Cal. Are you changing that fast?"

"What did Anika tell you?"

Even before he asked the question, Cal felt himself bifurcate: there was the Cal who wanted to whisper Frida's name into her hair until it was like he wasn't saying anything at all. And there was this new second self. He wanted to hear what Frida had to say, and he would weigh what her story meant, both for their family and for the Land. The other men seemed suspicious of Anika, and maybe they were right to be. What was she telling his wife?

Frida hadn't answered. She was squinting at him, as if trying to see something far away. Maybe she'd seen that he'd broken into two.

"I know you miss Plank," she said, "but you can't pretend like you're back there. This isn't college, Cal. It's not about harvesting vegetables and reading boring books all day."

"I know that," he said. "That's like me telling you this isn't the Ellis Family Christmas."

"Fuck you," she said.

"That's nice. Thanks."

"Cal, my brother chopped off a man's head."

"What? Whose?" His insides spun.

"A Pirate's. He killed a Pirate and chopped off his head in front of the children."

"What children?"

Frida let out a little moan, like she was imagining the worst and couldn't bear it.

"Frida? You okay?"

She shook her head. "I want to stay here, too, Cal. What other choice do we have? But we need to know everything. Find out, okay? Find out everything." She paused. "Even if you can't tell me, just find out."

"Next time you're scared, come to me."

Frida laughed. "Sure, okay. If you let me."

"Don't be like that." And then he said, "You find out everything, too. No more secrets."

Frida didn't agree or disagree. Instead she said, "Don't get lost in the dark tonight," and stepped out of the Bath, leaving without a goodbye.

———

After his nap, Cal had done his best to avoid Frida before his security duty. He hoped that by the time he was scheduled to return from his shift, she might already be with Anika, baking dinner rolls or pumpernickel or some elaborate tiered cake. He didn't want to see her because he didn't know if she was angry with him, or still upset, or what, and he'd never been so confused about his own wife. He certainly didn't want to rehash their conversation from earlier.

Sailor and Dave were waiting for him on the porch of the

Hotel. Other men were on the detail, Dave explained, but at various locations: on the second lookout Tower, patrolling the Forms, and so on. Frida wasn't entirely correct—a couple of women did participate, but never in the Towers or the Forms and never with a weapon; they sat on porches or walked the center path. If they fell asleep, no one cared; the other morning Micah had actually said, "If a lady dozes off, let her." If Frida knew her brother had made such a comment, she'd be livid.

But, no, Cal thought, he wouldn't let Frida into his mind. He'd focus on security. It was a welcome distraction.

Cal followed the boys toward the Bath.

"Our job is to watch for anything out of the ordinary," Sailor said. "It rarely happens—in the whole time I've been here, only a few people have ventured into the Forms. And they quickly turn back if we exert enough pressure."

"What do you mean by 'pressure'?"

"If you see something," Dave said, "we want you to blow your whistle. Whoever's been designated the muscle will go investigate. The rest of us wait as backup."

Sailor smiled. "I was the muscle the day you and Frida got here."

"We'll spend some time in the western Tower," Dave said, "and then we'll separate for an hour or two."

"We need to show you the lay of the land first," Sailor said.

There were a few people walking the main path and sitting out on porches, bundled in heavy coats and blankets. The Hotel and the Bath anchored each end of this path, and it reminded Cal of a promenade. From across the field, right by the barn, smoke rose from one of the two nightly campfires where people congregated to play music and tell stories. Peter told Cal that he made it a point of attending at least twice a week: "Just because we're in charge, doesn't mean we should live separately," he'd explained.

The houses had all been winterized, their windows giving away nothing. Cal wondered which rooms were filled with people, and which were still empty. Right now somebody must be pulling up the covers and blowing out a candle.

It was cold out tonight. His breath would be visible by midnight, but with a hat on, and his sweatshirt, Cal would be comfortable. The sky was obscene with stars.

Once they'd passed the Bath, Dave said, "Here." He handed Cal a whistle on a string.

"Put this on," he said. "We'll go over the calls in a few minutes."

Cal nodded. The whistle felt good around his neck.

"Also this," Dave said. He was giving Cal his gun back.

"I cleaned it," Sailor said. He grinned as he passed Cal a small box. "More bullets."

Cal nodded, clutching the gun. He tucked it into the back of his pants and put the bullets in his pocket.

"Ours are in the Tower," Sailor said. Guns, Cal realized.

Cal had always wondered what it might be like to climb a water tower. The ladder, with those semicircles of metal jutting out every few rungs, as if they might keep anyone safe. Each platform was another dare. *Are you sure you want to climb higher?* As a kid, Cal wondered how scary it might be to reach the top. Did the whole thing rattle in the wind, sing like shaky old bridges did?

This Tower was built of splintery wood, and its sawdust smell made him think of fall carnivals and pumpkin patches. The ladder reminded him not so much of a water tower but of the high dive at the local pool his mom would take him to during the summer. Damp. No frills. Possibly unsafe. But once you were up there, you definitely couldn't turn back, or you'd run the risk of embarrassing yourself in front of the kids below. He understood why Dave and Sailor had told him to go first.

Up, up, up he went. When Dave started climbing, the whole tower seemed to sag with the added weight. Up, up.

When he got to the top, he didn't look over the edge. Not yet. The Tower's room was a small turret, with walls that went chest high. The floor was crowded with a bucket of rifles, a pile of coats, some miners' helmets with flashlights attached, a megaphone, and a bedpan. Empty, thank God. From a hook hung the binoculars he'd seen Peter and Dave using the first time he'd caught sight of them.

Once Sailor and Dave had reached the top, Cal finally allowed himself to look out. The moon had been full the night before, so there was some light. From here, the Land seemed almost puny. Just that one strip of buildings, and the field with its barn, garden, and showers. Encircling it all were the Forms. From here, Cal could see where they ended.

Beyond all this: trees and more trees. They were tall, and in the dark their greenery turned woolly. To the east, a speck of orange firelight pulsed. Cal blinked and saw, farther out, a second fire.

"There," Sailor said, pointing north, "the old highway cuts through."

"Where does it lead?" Cal asked. He suddenly felt vulnerable. He didn't know what was beyond the Land, and he didn't know if he could trust the people who did.

Dave smirked and said they better go over the whistle calls.

If Cal saw anything suspicious in the Forms, he was to blow one long whistle. "Do it until you've run out of breath. Your job on security is simple: to watch for outsiders and to alert us if anyone tries to enter the Forms. You're patrolling for anything out of the ordinary. And I mean anything. If we end up going nuts about a rogue raccoon, so be it."

"We also have arm signals for the daytime," Sailor said, "but they won't do you any good while it's dark."

"For tonight you only really need that one call," Dave said. "The others you can learn later."

As soon as the lesson ended, Dave walked to the other side of the platform, scanning with the binoculars. He pulled a walkie-talkie out of his jacket and began exchanging observations with someone in the Forms. The reception was scratchy, and Cal didn't recognize the person's voice. He had only just learned of the walkie-talkies that morning. August was going to lobby for a few more sets from Pines, plus more batteries.

What wouldn't they ask for?

"No women ever want to come up here, and do this job?" Cal asked. "I mean, come on, everyone digs walkie-talkies."

Sailor shrugged. "You assume this is a superior occupation." He nodded at Dave, who was now watching the landscape. He looked like a dog, waiting for its owner to come home.

"Anything withheld long enough does start to seem better, don't you think?" Cal said.

"You should write fortune cookies," Sailor said.

Dave hushed them. "Guys. Focus."

They fell silent and watched for movement as Dave had instructed. Cal kept his eyes roaming as Sailor explained that some of the Forms had been there long before anyone on the Land showed up. Cal held his breath as Sailor talked; he wondered if Frida had heard this before or if she'd learned a different history.

"But we built a whole lot more, and we designed it so that they'd form the maze you walked through," Sailor explained. From above, they were spirals. "Some of the spirals are square shaped," Sailor said. "Come up during the day, and you'll see.

There *is* order to it. And the glass in the ground? That was my idea. Saw it in Peru when I was six years old."

"What were you doing there?" Cal had never met anyone who had been out of the country.

"Guys," Dave said again, and Sailor didn't answer.

After what felt like an hour, but could've easily been twenty minutes, Sailor said, "Break."

Dave continued to watch, but Sailor nodded at Cal, and the two men slid to sitting in what little free space there was.

"We take forty-minute shifts," Sailor explained. "But Dave always starts out crazy."

Sailor brought out a rag from his coat pocket, and he untied it to reveal a handful of pumpkin seeds. He gestured to Cal to take some.

Cal took a few of the seeds. They weren't coated in salt, as he'd hoped, but bare, a couple of them still slimy with pumpkin innards. "Can I ask you something?"

Sailor waited.

"How did you end up here? Why did you come to the Land?"

"Why do you want to know?" This was from Dave. He held the binoculars to his face.

"I was just curious," Cal said. "I mean, why here?"

"We were recruited," Sailor said. "By Catherine with a *C.* We called her Catie."

"Catie with a *C,*" Dave said with a little laugh. "She was awesome."

"Are you sure that was her real name?" Cal asked. "I mean, I'm thinking it could be the same person who recruited Micah. Toni? Short for Antonia."

"I doubt it," Dave said. "It was Catie's first year doing it."

"They used a different recruiter every year," Sailor said. "Cleaner that way."

"'Cleaner'?"

"As in harder to trace back," Dave said. "The same woman coming to visit a bunch of boys at a weird school, year after year? That's bound to draw attention eventually."

"So Catie with a *C* was part of an ongoing practice, to get Plankers?"

"You bet," Sailor said.

"What did she tell you?" Cal asked.

"What did Toni-short-for-Antonia tell *you*?" Dave asked.

"Nothing," Cal replied. "It was Micah she wanted."

"Then she wasn't a good recruiter," Sailor said. "They're supposed to get at least two or three men interested."

"Why Plank, though?"

"Why not Plank?" Sailor asked. "We know how to farm and how to cook and how to build shit. You need all that in this world. Plus, we're intellectually curious."

Dave laughed. "I guess they tried to get some off-the-land types first, but they weren't focused enough or smart enough or were just too hard to find. They needed people with skills, and Plankers have them."

"You're talking about the Group, right?"

No answer.

He tried again: "Sailor, you said not everyone on the Land participates in the Group's activities?"

That's when Dave turned, binoculars falling around his neck. "Jesus, Sail. You're worse than a teenage girl."

"He's at the meetings," Sailor said. "What difference does it make if he knows?"

"Knows what?" Cal asked. *Find out everything.* "What did the recruiter tell you?"

"It was different for Micah," Sailor said. "He'd never heard of the Group when he was in school. But by the time we were at Plank, we'd read about the stunts and Micah Ellis, the infamous suicide bomber, and even the encampment. The school was a month from closing when Catie showed up."

"God, that was depressing," Dave said. "Remember how classes kept getting canceled?"

"She talked to us about the projects the Group was undertaking to revitalize L.A., and she gave us books to read. She told us what similar organizations were doing in other cities." Sailor paused. "I seriously had no idea where I was going to go until she came around."

"But you didn't go to a city," Cal said. "This is nowhere."

"We trained in L.A. first," Sailor said. "The Group wanted to end senseless violence in areas with minimal population."

"Can you be more specific?"

Dave raised an eyebrow. "Pirates, Cal. Bands of marauders. Sailor is babbling about how we were trained to protect settlers in this area from Pirates."

Sailor's voice pitched a little higher; he was excited. "The argument was, why should those criminals take and take from innocent people? They had no mission beyond greed. In some ways, they were a lot like the people who started the Communities, taking from the less fortunate, hoarding it for themselves. We came with about twenty others to make this area peaceful."

Cal couldn't find the words to speak, and Dave smiled. "That's why you're still alive, my friend. Because we eradicated the threat."

"You're welcome," Sailor said.

Dave sighed. "Break."

Sailor rolled his eyes, and stood.

Dave sat across from Cal on the floor. Cal reminded himself to be quiet, let them keep talking, let them lead him to the answers. *The meek will inherit the earth:* wasn't that the phrase? He wasn't even sure what that meant, even after the Sociology of Thought seminar at Plank. During that class, he'd almost wished his mother had sent him to Sunday school. The Christians, and the former Christians, they'd had a leg up during the lectures on Jesus.

Dave yawned loudly and Sailor, eyes to the binoculars, said, "The more Dave yawns, the better he is at this job."

"Good thing yawns are contagious, then," Cal said.

Dave laughed. "You're witty, California."

Cal winced. He'd noticed, over the last week, how every person he got close to on the Land turned into Micah, if only for a second. It could be a small, borrowed gesture or some phrase Micah had used the day before or his method of skating across an insult in such a way that it was barely detectable to the victim. Back at Plank, Cal had done it, too. Once, in a class, he'd found himself drawing goats with human faces in the margins in his notebook, the very same beasts Micah liked to doodle when he was bored. Cal had stopped midscrawl and even gone back to black out the evidence of his mimicry. Was he ashamed that he'd imitated his friend so brazenly? Or was it that he knew, in his heart, that Micah was inimitable?

Fuck it, Cal thought. The meek didn't inherit a thing. "Why would the Group want boys from Plank?" he asked. "So what if we can grow food and talk about Kant? That's hardly what makes someone want to join the Group. It doesn't compute."

Dave laughed. "Wow, really? That's like saying you don't see how so many radicals could come out of Berkeley in the 1960s. These things happen. I hate to be the one to tell you this, but the Group was coming to Plank long before you got there, and long after you left."

Sailor nodded. "Why do you think it was still open when so many other schools were shutting down?"

"But not everyone there was a terrorist," Cal said.

"Of course not," Sailor said. "I'm not a terrorist, either."

"But you're in the Group," Cal said.

"It's not a gym membership," Dave said. "It's not like they give you an ID card when you join."

"Who knows what we are now anyway?" Sailor said.

"What does that mean?" Cal said.

"Back at Plank," Dave said, "Catie told us about a frontier that needed to be tamed. She said the Pirates were mercenary killers who needed to be stopped."

"And that appealed to you, even though you'd be connected to the Group? You knew what they'd done, you just said so."

"Eventually we got to meet Micah, work with him," Dave said.

Cal said, "I wouldn't have pinned you for a killer, Sailor."

"There's more to it than that," Sailor said. "I did what was right."

Cal remembered what Frida had told him. If it had been Pirates that Micah had killed, wasn't that a good thing? Not killed, *but beheaded*.

"We helped get this place settled," Sailor said.

"What about the people who were here before you guys showed up?"

"They mean well," Dave said.

"Do they know what's going on? That you're part of the Group?"

Someone was coming in on the walkie-talkie. "All clear?" Dave said, and waited for the person to say, "Affirmative."

"Ask Mikey," Dave said to Cal.

"Most people on the Land live simple, happy lives," Sailor

said. "They do their chores and make food together. It's a peaceful place."

"Are they mistaken?" Cal asked.

Dave shook his head. "Of course not. It's just that their lifestyle needs to be supported by a select few of us. We're active, so they can be passive."

"'Active'?"

Cal said. "Ask Mikey," Sailor said again. "He'll explain."

"We don't know that, Sail," Dave said. "Break."

A few hours later, after Cal had assiduously studied the hand-drawn maps Sailor had given him, he headed for the Forms. Dave had insisted he wear a scarf. "Before dawn is when it gets the coldest, and if you get stuck, you might be out here for a while."

"The trick is to compare what's in front of you with what you remember from the diagrams," Sailor said before they parted ways.

"Do you have that kind of mind?" Dave asked.

"I guess we'll find out," Cal said, and clicked the light on the helmet he'd been given.

Beneath the moon, the Forms had been eerie, but in the funnel of light they were sinister. The objects within them had been wrested from their original purpose, and they were now misunderstood, lost. It must have been a huge sacrifice, Cal thought, for the residents of the Land to give up these things in the name of protection.

He noticed an empty picture frame and then the boomerang shape of a rocking-chair leg. A doll speared with barbed wire. He took a deep breath and reminded himself that they were just objects. He forced himself to look again, and closely. "Commit all this to memory," Dave had said before leaving him alone.

Cal had no exact purpose except to walk through this maze with an alert mind. He had his pistol and the walkie-talkie, and Dave would find him should he need help. The goal, really, was to just get used to the Forms, to learn them as he would streets in a strange city.

He walked slowly, careful of the glass at the outskirts. The easiest way to understand the layout of the Forms was to imagine a series of spirals. He studied each Form as he passed, memorizing its contents. The door of a washing machine; a plastic air-freshener plug-in; a shopping basket missing its handles. Just objects. Collected by people, reused by them.

Not many had seen the maps. Sailor had unrolled them solemnly, said, "These are closely guarded," and hovered as Cal read them. Cal understood that the others—the ones not in the Group—were hemmed in by these Forms. They couldn't leave easily, not without people finding out. Not that they wanted to leave. They lived a good life here, and that was all anyone could hope for nowadays.

Cal stopped walking and turned off the lamp on his helmet. The miner's cone of light disappeared. His eyes should get used to the dark, as an animal's did, and as his own had, back when he and Frida were holed up in the shed. He imagined the Form next to him exhaling, relieved to be back in the safety of darkness, and Cal felt a kinship with the thing.

In L.A., when Cal could no longer improve the world by growing food, he had resolved to escape the wretchedness, take Frida to the edge of the world, and start over. And when he discovered the Millers had killed themselves, Cal had buried their bodies and resolved to retreat once more from ugliness, from the familiar wretchedness that seemed to follow them. How impossible, though, to turn one's back on all the horrors in the world; there had to be another way to live.

Passing beneath these makeshift monsters, Cal understood that he was now part of Micah's scheme. He would be man enough to admit it. Frida was wrong; he didn't think this was Plank, the sequel. This was a brutal wilderness where people did what they had to in order to survive. By taking back his gun and ascending the Tower and studying the maps, Cal had tacitly accepted whatever was going on behind closed doors on the Land. That's what Frida would say. She was right about that, but maybe that wasn't a terrible thing. Cal had rejected the Group back in L.A., but out here, where there were Pirates to fight, and people to protect, it was easier to accept. It made sense. There had to be a part of Frida who agreed with that.

The past didn't matter if it was the future they had to worry about. It would all be okay; he'd have to convince Frida of that. At least for now, alone in the dark save for the glow of the moon, it seemed okay to him.

Cal soon found that he wasn't lost, that he could recall the maps Sailor had unrolled on the tower floor and see the maze's tricks before him. He did have that kind of mind. He walked carefully but with confidence. He put a hand out to the closest Form, gentle enough that the edge didn't cut him. They were just objects. Made by people to fight people, and to protect them, too.

Maybe all along he had wanted Micah to find the revolutionary in him. He'd wanted someone to seek out that place in him that could be powerful.

17

It was darker than dark, and far too early to meet Anika downstairs, but Frida couldn't stay in bed any longer. Cal wasn't back from security yet, and thank goodness, because he was the whole reason she couldn't sleep. Imagining him out there, patrolling the Forms, or watching from the Tower for any suspicious movement, made her sick.

She crept out of bed and put on her shoes and coat in the dark. She lit a candle, and followed its light out of the bedroom and down the hallway. The stairs creaked with each step, but she kept going, one hand dusting the banister as she went.

Once Frida was outside in the cold, she followed the flame's flickering light down the path. If Cal was up in a Tower, he'd see her. Was that what she wanted? She couldn't shake the thought that he might blow his whistle instead of climbing down to talk. *Stop being so dramatic,* she reminded herself. She'd asked him to find out everything, and that's what he was doing by watching the borders until sunrise. He'd learn this place for the both of them.

Frida let the candlelight dance across the dirt. She kept her eyes on the flame but didn't move. She had no plan, nowhere to go. Every time she thought of her fight with Cal, and about Anika's story, about the Pirates attacking this place, about the kids who used to live here, about her brother beheading another man, and about Cal being blind to what had happened here, she wanted to scream. She stared into the flame. It was so feeble against the blackness.

She always took off when they argued, as if it might kill her to stay another second with Cal when she was angry. She should have stayed, shouldn't have been so gutless, so afraid to hear him out. He wanted to make things better, no matter what it took, and she couldn't stand that.

The Land didn't get as quiet as the wilderness did. Even in the middle of the night, Frida could hear two people talking nearby, and in another house, someone was humming. There was the occasional sleep-snort, too, and the creak of a bed as someone rolled over. There was so much life here.

Frida shielded the candle with a cupped hand and stepped forward, one foot in front of the other, as if walking on a balance beam.

Before she'd left the Bath, she had asked Cal to find out everything. He'd asked the same of her, but she'd nearly forgotten that part. It was her job, too, to discover the Land's history. Cal said Micah didn't care about the past, but that had to be false. Everyone cared about the past.

Frida had never been inside Micah's house, but it had been pointed out to her a few times, even by her brother himself.

"If you ever need anything," he'd said. "That's where I sleep most nights."

"'Most nights'?" she'd asked, an eyebrow raised.

"Not like that," he said. "I don't have a girlfriend if that's what you're getting at."

It was a one-story structure made of wood that had darkened and gone brittle over the years. Its roof sloped, and its two front picture windows were boarded up.

Across the doorway hung a curtain made of thick material

303

and torn at the side. It was probably as cold in there as it was out here, Frida thought, and dark. As she stepped up to its entrance, she practiced what she would say: *Tell me exactly what went on here. What have you done in the name of containment, and where does it stop?* She would ask him what happened to the children. Anika shouldn't have to be the one to retell such horrors.

"Hello?" she said as she reached the doorway. She pulled the curtain aside, prepared to walk into the house, when her hand hit something solid. Frida saw that behind the curtain was a wooden door, sturdy as the Hotel's, sturdier than Anika's, and a metal knob. Of course her brother had a door, it was just hidden.

Frida struck her knuckles against the wood, hard, and then once more. When there was no answer, she turned the knob. With a click, the door opened.

She found herself inside one large room. Two windows across from the door would let in the first of the morning's light; they had not been boarded up, Frida realized, because their glass was intact.

A shelf ran along the length of the room, directly below the windows: it was smooth, and when Frida shined the candle she saw it was made of wood. Different candles lined this shelf, some of them burned almost to nothing, others long and tapered. None of them were lit, but Frida imagined that when they were, they transformed the room. It'd be like standing in an elegant little restaurant: spare, honey lit. All it lacked was a hostess stand, some skinny cute woman with a handful of menus, ready to show you to your table.

Against one wall was a single sleeping pallet, empty, and in the center of the room a stove huffed. It was warm in here. Micah must have left recently.

Frida was stunned by the anger she felt. Or maybe it was

envy. The house's exterior was nothing special, but it was welcoming inside, almost beautiful. Someone had renovated this carefully, but if you saw it from the outside, you'd never have any idea. Everyone knew Micah lived here, but how many were invited inside?

She stepped to the window and for a moment felt like she'd found herself in a charming country home. That didn't seem like Micah, though: he'd want to wake up to see the Land's dangerous and unique border. Maybe when the sun rose, it revealed a line of sharp Spikes in the distance.

Frida pushed away from the window, the glass solid and smooth against her hands, and rounded the stove. Behind it was a table, low to the ground, and beneath it, a pile of clothes, a few folded neatly, others flung carelessly aside. Frida picked up a blue T-shirt and held it to her face. There it was, her brother's smell, as if he were still fourteen years old, showering every morning for three minutes, timed, like he was training for the military. Frida knew she was caught in a fantasy, but she didn't care. She breathed in deeper.

On the table were a few odds and ends: a fingernail clipper, a brush for Micah's long hair, and a bandage, the kind you'd roll onto a sprained ankle.

And then she saw the toy.

They had called it the Bee, even though someone, Dada maybe, eventually realized it was a butterfly. By then, it was too late, the Bee was the Bee.

It was a plastic butterfly with clear blue wings and a big smiling face. From its head protruded a ring that opened and closed; Hilda used to hook it onto their stroller or onto one of their car seats. Not that Frida remembered any of that; when they were older, the toy used to sit on the mantel like a vase, and Hilda would sometimes talk about it. When Frida and Micah were babies, she said, the Bee had the miraculous power to turn their distress into something more palatable. It had saved the family on many crosstown car trips.

Frida picked it up and rubbed her hand over the ridges of the wings, across its smile, its big orange eyes. Its body was striped, black and white, but now the white paint had peeled off, revealing a sad gray color beneath. The Bee.

Her brother had taken this toy from their home. He must have wanted a souvenir before he left L.A. for good, and he knew it wouldn't be missed. Frida had forgotten all about it.

There was no way Micah saw this now and didn't think of their family, of their mother's stories. He might not want children on the Land, but it wasn't because he was evil. He had to have a reason; it had to be an act of compassion.

Maybe her brother would give this to her, pass it on to his niece or nephew. He loved her, and he loved his family. Frida remembered what he'd said to her in the tree house. He thought she'd go to the encampment with their parents. He'd wanted her to be safe, too. Her little brother was mixed up, but she could forgive him for that.

———

As Frida walked into the kitchen, Anika said, "I thought we'd try breadsticks today."

The kitchen was dark, the candle between them throwing shadows across Anika's face.

"Those were my brother's favorite, growing up," Frida said.

"Oh, yeah?"

"He loved stale, store-bought ones. He loved how crunchy they were." Frida laughed. "It was kind of weird, actually. But also funny."

"Sounds about right," she said.

As they worked, Frida said, "The kids had to leave when Micah got here, right? That was a condition of his help?"

Anika nodded, but she didn't look up. "It was practical."

"I thought so," Frida said.

"August had access to a Community. At first, we didn't know which one."

"And you didn't know it was Pines until August brought the objects for the Forms?"

Anika didn't answer.

"Anika?"

"Back then, the Communities had everything figured out except one thing: children. What if someone couldn't conceive, even after IVF and all that? What then? It didn't happen often, but occasionally, there was one unlucky couple on the block."

Frida remembered what Toni had told her: that in Communities, childless couples were frowned upon.

"August took the youngest children there," Anika said. "To live."

"Adoption?"

"They wanted babies."

"What about the older kids?" Frida asked.

"Bo and Sandy took Jane to live off the Land. They were the only ones. A handful of others were going to do the same, but right before the Millers left, another family lost their child. Melissa—our oldest. I told you she had that fever when the Pirates came? Well, she died of it."

"My God."

"It frightened us, the thought of being out there alone, vulnerable to something like that happening. Micah told us we were right to worry. I'm not sure if his goal was to scare us into keeping close, but it worked. Melissa's parents are still here."

Anika didn't say their names, and Frida didn't ask. She'd let them be themselves, not their tragedy.

"I assume Pines was able to take care of the children?" Frida asked. "They wouldn't die of fever there. But I bet it's terrible sometimes being without Ogden."

Anika nodded. "It is. But I'm glad he was young enough to be adopted. Micah said he'd go to a family immediately, a well-off one. The older kids didn't have it as easy."

"They weren't adopted?"

"No. They were sent to a place called C.A.P., the Center at Pines. That's where children too old to be adopted are educated and trained for jobs. Micah said they'll be well fed and safe. He showed us pamphlets. It looks like a nice boarding school, with classrooms and a cafeteria. Once they're old enough, they receive apprenticeships. The kids who grow up at C.A.P. are guaranteed jobs upon turning eighteen. It's manual-labor stuff. They'll only be eligible for certain jobs, but they won't die of fever or starvation, and their lives will be easier there." She paused. "Micah said it was the best thing for them. Didn't we want a better life for our children?"

"But to give them up—" Frida stopped herself midsentence. She wished she could take it back.

"I'll never forget it. All the kids left together on the school bus, dressed in clothes August had provided for them. Crisp, clean dresses for the girls and pants and button-down shirts for the boys. Even tiny outfits for the little ones. Ogden had a smile on his face when they put him into his carrier, like he was proud of how he looked, like he was excited for the ride. We could almost pretend it was normal, our babies' first day of school. We waved until the taillights disappeared."

Frida remembered the bus, parked in that meadow like something out of a children's book. Frida imagined Anika giving her baby away. He would be covered in a light blue blanket, to protect against the chill of the early morning, his tiny clenched fists hidden beneath it. Had Anika run her index finger over Ogden's gums one last time, to feel the teeth cutting through? Did she cry out as Ogden's familiar weight left her arms? Or did she remain stoic? As something dark pressed at the edges of her chest, did she press back? This was best for Ogden, she must have told herself. Wasn't it?

The day they found out Micah was dead, Hilda said she could still feel the top of his baby head beneath her nose, against her mouth. She said she remembered the way she'd comb her fingers through the fuzz of his hair as she nursed.

Anika gestured for Frida to step aside so she could roll out the dough.

"You can't think too hard about this, Frida. We all had to make sacrifices. I suppose that includes your brother as well."

"Micah is sweeter than he lets on," Frida said with a smile. "You know he kept this toy? We loved it when we were kids. It's a little bee, well, it's a butterfly. But my brother has it in his room."

"The Bee?" Anika said.

"You know it?"

"He gave that to Ogden, before they left." She put both hands on the table before her, to settle herself. "He took it back?"

"I'm sure there's an explanation . . ." Frida didn't know what else she could say. Her brother had given the toy to a baby, and then he'd taken it away. It was petty at best. At worst—she couldn't go there.

She wondered what other cruel things Micah was capable of. She tried to imagine Randy hanging the Pirate's head from the top of the tallest Spike. He was probably crying, and her brother would have remained calm, as if instructing the boy how to decorate a Christmas tree.

"What about Randy?" Frida asked.

"He's at C.A.P. now, too."

"Deborah let him go?"

"It was the only way," Anika said.

They didn't speak for a moment, and then Anika placed both her hands flat on the table and said, "Frida, let me tell you about your brother."

"I know my brother."

"You don't," she said.

Frida waited.

"The first year Micah was here, I was very difficult to live with. Losing my baby was harder on me than the others, I don't know why. I wasn't very cooperative, I talked back, I didn't want to help with the Forms, or anything, really."

Frida wished she could stop listening; she knew something bad was coming. If only there was a door to slam, a bridge to jump off. But she let Anika keep talking.

"On a particularly dark day," Anika said, "I refused to show up to the Church's meeting. I lay in bed all day, crying. I thought I'd be reprimanded publicly, but it was worse." Anika stopped.

"What is it? Just tell me."

"Micah came to my room when the others were outside working. He told me that if I didn't get in line, there would be no place for me on the Land. He said it would be worse than I could ever imagine. He was whispering. He said the Pirates were still out there, beyond the Forms we were building, and that they'd kill me if they ever got the chance."

Frida didn't say anything.

"I didn't talk back, I just wanted him to leave my room. But he didn't. Instead he leaned close and whispered, 'Ogden was yours, right?' He began to describe my son: the color of his eyes, the birthmark on his left arm, the shirt he was wearing when August carried him onto the bus. I started to cry—how dare Micah threaten me like that? I'd already given him everything."

Anika was crying now. She wiped her eyes with the back of her arms. "Then he turned and walked away. I had my eyes on his back.

That's when I saw the red peeking out of the back pocket of his jeans. It was just the edge of a bandanna, grazing the hem of his shirt, but as soon as I saw it, I felt that same jolt of fear, and I had to shut my eyes. I should have screamed. I don't know why I didn't. Micah had put the bandanna there, so I'd see it, I'm sure of it. He must have heard me gasp because he turned around once more. He actually smiled at me.

"He said Ogden would be safe. All I had to do was attend the meetings and get along with everyone else. With him."

"And so you did?"

"First I tried to tell Peter what had happened. He said I was overreacting, that Micah was just trying to get me to cooperate. Peter didn't understand. He said he missed Ogden but had always been afraid for him. Now he wasn't. He thought I was mistaken about the red bandanna, but I wasn't, I just couldn't prove it. And then Micah announced to everyone that I'd be moving to a bedroom with a door and that I'd get to take over the kitchen. He was giving me special privileges like a bribe for good behavior."

"That's one way to look at it," Frida said. "Or maybe he was just trying to apologize and make things right after he scared you so badly."

Anika pushed the heels of her hands against her eyes. When she removed them, her eyes were pink, her face drawn and tired. "Maybe you're right. Maybe Micah was being smart, and he knew it wouldn't take much to bribe me into submission." She shook her head. "Your brother, he swept in, over us all, and we couldn't stop him."

As they finished baking, Frida kept replaying Anika's story about her brother and the children in her mind. There was no way Cal had any idea what Micah had done; if he had, he would have told her. He'd be too afraid for their own unborn child not to.

At dinner last night, Frida had noticed again how young Sailor looked. Dave, too. And Burke. And the guy she'd waved to on the way back from the shower, Doug. These men were barely out of childhood. The mothers on the Land must have sensed this and welcomed them happily.

Frida was probably the youngest woman on the Land. Fatima had to be in her midforties; Sheryl, too. Pregnancies could happen, but accidents wouldn't be likely. Kids had been removed from the future.

Perhaps Anika had done the right thing and given Ogden a safe upbringing inside of Pines. And maybe it was easy to give up parenthood for the life Micah offered them. A life of regular meals, warm showers, leisurely afternoons, and a bed. Cookies when they behaved. Maybe the one thing that every parent wants is to be a child again, to be taken care of.

Micah had planned everything, Frida realized. He had foreseen every motivation.

She imagined her own child, small as a fig. He would be good. He would be necessary. Her brother would see that, and he would accept him.

He had to. If not, then what?

Frida wanted to believe that her brother had sent those kids to Pines because he thought they'd be better off. But there had to be more to it. Micah wasn't selfless. He had to benefit in some way.

Frida was at the trough rinsing the baking sheets when she heard someone whisper hello at the front of the kitchen.

There was Fatima, waving. Anika turned to Frida with a tight smile.

"Good morning, ladies," Fatima said. "Micah said I could join your baking lessons. He thought you might need some help."

18

Sailor and Dave sometimes missed the meeting the morning after a patrol, and they told Cal he could sleep through it today, too. August had left, and only Peter and Micah remained. But once Cal got into bed, he was restless, and he couldn't even close his eyes. He was thinking about Frida, about what he'd tell her about the night. He wanted to describe how the Forms shined in the moonlight, how magical they were, and how it felt to see the Land from way up high in the Towers. He thought, maybe, she might understand. She'd said it herself: she wanted to stay on the Land; there was no other choice. Maybe they weren't fighting after all.

He couldn't interrupt her in the kitchen, though, not without causing a scene. The last thing he and Frida needed was to draw attention to themselves. He might as well do as he normally did. He got out of bed and put his boots back on.

Inside the Church, Micah lay curled into a pew, his arms crossed, his chin to chest. He didn't move when Cal walked in; he was asleep. Goose bumps had spread across Micah's arms, and his legs looked tense, as if they were trying to kick off the cold.

"Micah," Cal said, and he jerked awake.

He saw Cal and let out a little laugh, then sat up and yawned. His hair hung stringy across his shoulders.

"I couldn't sleep," Cal said.

"I was dreaming."

"You sleep here?"

"Sometimes."

"Without a blanket?"

Micah shrugged. "A lesson in fortitude." He stood up and stretched. "Fuck, it's cold. Good thing you made it out of the Forms. Dave is still impressed. So is Sailor, but, come on, he's like a toddler at the aquarium. You're his jellyfish."

"Micah, they've told me some things."

"I know."

"These past few days, I've been thinking."

"I noticed." He raised an eyebrow. "You really want to know how this place works, don't you?"

"I have to. We can't be in the dark anymore."

"'We'? I told you, what you learn in the meetings is secret."

"I was on security; it wasn't a meeting."

"It's still sensitive information."

"Is it? We talked mostly about Plank, which, we all know, only Plankers care to hear about. Dave and Sailor told me Toni was probably a bad recruiter."

Micah smiled. "They're right. She fell in love with me that weekend we met, and everything went downhill from there." He blew on his hands before rubbing them together for warmth. "Though she did enjoy being the only woman in a sea of men."

"More like a pond of boys."

"More like pond scum."

Cal laughed. "And you liked that she chose you."

"Are you kidding? I loved it. It's probably why I got so involved.

I'd been singled out, made to feel like I was destined, like the

Sun King." He paused. "And it's probably why I was so devoted to her. At first."

"'At first'? You always said you were totally faithful, that she was crazy for being so jealous."

"She was, but that doesn't mean she was wrong. I did have a wandering eye. She simply couldn't hold my attention."

"You're a dick."

"Tell me something I don't know," Micah said.

"Where's Toni now?" Cal asked.

"She's not in the Group anymore, you knew that."

"Are you?"

"Yes and no."

"That's not an answer."

Micah stood up and stretched. Cal waited for him to finish.

"Micah?"

"When the Plank boys first joined us, I asked them about the farmhouse. I don't like nostalgia, it's useless thinking, but I found myself missing those years, wanting to bring some of that time back. Not that I could ever do that, not really."

"I can't blame you for trying," Cal said. "It was beautiful there."

Micah closed his eyes, and for a moment Cal saw him as his old roommate. Micah's cluttered desk and unmade bed, the socks he'd wear two days in a row before changing into a new pair. They'd stay up so late some nights, reading, drinking, talking—about what? T-ball. Adorno. Whose grandparents were weirder. What if they'd been born in 1472. Or 1981. Or 2015. Or tomorrow: "Abort me, Mama" was all Micah would say about that hypothetical. After nights like those, they'd be yanked awake just an hour or two later by sunlight pouring into the room. One time, Micah groaned and threw a sneaker at the window; the shoe bounced off the glass, thankfully.

Neither put up a sheet to cover the light or even discussed it. That wasn't how they did things at Plank.

"Come on," Micah said now, opening his eyes. "I want to show you something."

Micah began walking toward the stage, and Cal followed. He no longer felt restless; he was calm, as if he'd slept deeply all night long.

"Where's Peter?" Cal asked.

"In bed," Micah said, without turning.

When Micah reached the door behind the pulpit, he lifted his pant leg and pulled a ring of keys from his boot.

"Better than a knife," Micah said, and turned to open the door.

The door had just one locked knob, but when Micah opened it, there was a second door, also locked.

"Two doors?" Cal said. "Wow."

Micah wrestled with the lock. "Came like this. It's mostly for show. Nothing that can't be bulldozed or blown up."

"You should know," Cal said. "Or, no, I guess not."

"Touché," Micah said. "But just because I wasn't killed by a bomb doesn't mean I don't know how to make one. Remember how the guys and I used to blow up the empty feed containers?"

"You totally freaked out the livestock."

Micah hooted. "That's about all. They weren't very powerful explosives."

Cal had to step out of the antechamber so there was room for Micah to pull open the door.

"Here we go," Micah said, and Cal peered over Micah's head to see a short, narrow staircase, carpeted with what looked like Astroturf after one too many minigolf games.

Cal wasn't sure what to expect. Were they headed to the war

room? He imagined more maps, maybe a wall of weapons—machetes, machine guns, and sparkling, sharp daggers. Bombs. Bricks of gold.

Micah Ellis as James Bond? *Oh come on, Cal,* he thought.

He took the stairs two at a time, just as Micah did, arms winged, hands not holding anything.

As Cal took in the room he caught himself feeling grateful. No Bond here.

There were books. Real ones, with spines that cracked, pages that you could fold over, underline, tear out, even. Most were hardcover; Cal hadn't seen those in years, not since he'd graduated from Plank. There was no way Micah would have shown this to Frida.

"Awesome, right?" Micah said.

Pushed against the opposite wall was a ratty couch made of crushed velvet so green it was yellow. *Chartreuse,* that was the word Frida would use. At one end of this couch someone had flung two gingham pillows, badly sewn, probably stuffed with the feathers of a sad, small bird.

On the table in front of the couch was a pile of comic books. Cal thought it was a series his father had collected.

Above them was a skylight, the glass still intact, the weak dawn light streaming through. The steeple's spire was visible through the glass.

"They put that in when they were rehabbing this place," Micah said, gesturing above. "It used to lead to the steeple, but now you can't access it that way. Not much forethought for authenticity, but at least it's warm in here."

"You're lucky the bell's not there anymore," Cal said. "Someone would want to climb up to ring it." A little boy, he thought. "You come up here a lot?" he asked.

"I do," Micah said, sitting down, "but not as much as I'd like."

He sighed and picked up a pillow. "It's silly, which is why it's a secret. It means too much to me to share."

"That's selfish."

Micah shrugged. "Sounds about right for me, don't you think?"

Cal laughed and walked to the bookcase. He resisted the urge to run his finger along the spines, but he read some of the titles. *The Prince. The Pleasure of the Text. The Waste Land. Bridget Jones's Diary. A Bereavement.*

"*A Bereavement*? Franzen's posthumous novel?" Cal asked.

"Some are from Plank."

"And the others?" Cal asked.

"The comics, they're from Burke's grandpa." He paused. "He has no idea I took them and put them here."

"You're an asshole."

"Again, sounds about right."

Cal sat next to Micah and grabbed the comic book. On its cover the superhero wore a mask and suit, its red and blue bisected with black lines, meant to look like spiderwebbing. He was climbing the side of a building.

"My dad used to read these."

"I know, Cal. We were roommates, remember? You used to go on and on about your dad."

"What does that mean?"

"Oh, nothing. Just that you didn't live with the guy, so of course he was godly."

Cal put down the comic.

"Sorry," Micah said.

It was the first time, as far as Cal could recall, that Micah had apologized. For anything.

"Micah," Cal said. There was opportunity here to find out the whole story. This was the moment he'd been waiting for, the moment both he and Frida needed. "If Frida's pregnant, what are we going to do? Will we be sent away?"

Micah groaned. "You and Peter . . ."

"Me and Peter what?"

"You won't let up. You want to know my plan."

"Do you have one?"

Micah raised an eyebrow. "In a way, yes. But it doesn't have anything to do with your baby. Who may or may not exist."

"He does."

"'He'?"

Cal took a deep breath. "I know the Land is opposed to expansion, to children."

"It makes sense, you know it does."

"Does it?"

"The Land was a mess when we first got here. There were children, but they weren't doing that well. Almost all of them were underweight. One had a skin infection that needed to be treated. Right after I got here, one girl died of a fever. A fever, Cal. Can you imagine? Almost all of them were still too young to contribute anything, and the adults spent a lot of time looking after them, and they couldn't get as much work done, couldn't make preparations for their own survival. That endangered the whole community. Plus, the older ones would be teenagers in a few years, and who knows what would happen then? They might not follow rules or do their jobs. Or they might decide to leave the Land and jeopardize everything." He paused. "Pines wanted children, and I could provide them with that. The kids are safe, and so are we. Everyone here agreed to the policy."

"And will that policy remain? The Land's different now."

"What I did wasn't an act of cruelty," Micah said.

"You can't send my child away."

"You're right, Frida wouldn't let me."

"*I* wouldn't," Cal said.

Micah said nothing.

"Haven't you considered passing all this on?" Cal waved a hand through the air. He meant the room, he realized. It was everything to his friend. Even after all that had happened, Micah was his friend.

"An heir?"

"Your word, not mine," Cal said.

Micah was trying not to smile. "You don't understand."

"Then make me," Cal said. "This morning meeting is just ten minutes in. Plenty of time left for you to tell me your crackpot ideas. Just like in our salad days, right?"

Cal expected him to laugh, but Micah had turned inward. When he looked up again, there was something fierce in his eyes, and Cal saw a man who was capable of murder, of beheading, of who knew what else.

"You think I'm just a shill for Pines," Micah said. "I wouldn't blame you, if you thought that. I mean, we work with them, so if you wanted to put it that way, you could."

"Do you put it that way?"

"Depends on who's asking."

Micah began to speak in a rush. It was as if Cal had merely reached over and turned a volume knob behind his brother-in-law's ear, as if Micah had been talking all this time, and Cal just hadn't heard him.

"I go about once or twice a year. August goes on his own the other times, as often as he can. Aside from bringing them information, we also trade fresh produce and cow's milk, that kind of thing. You knew that. Sometimes we bring them fish from the stream or barrels of our soil, which I guess is really something else. If you asked anyone on the Land, they'd tell you that Pines loves what we have to offer—the shit we make is *artisanal*."

"So—what then? August walks into Pines with a bucket of dirt?"

"I wish. The journey is hard, with the state the roads are in, but it's not impossible once you know the trouble spots. We take the bus.

Did you see it, on your trek here?"

Cal nodded. He remembered the pristine school bus.

"We have permits, which are updated at each visit, contingent upon our behavior, our information. There are more people allowed into the Communities than you think. It's pretty easy to fill out the paperwork." He waited for Cal to say something, perhaps to express surprise, but Cal said nothing.

"Communities have to communicate with one another," Micah continued. "It's obvious, I guess, but I never thought about it until someone explained it to me. It's more efficient for these guys to work together. Sometimes, at least."

Cal only spoke because he could tell Micah wanted questions to answer. "Work together on what?" he asked.

"For starters, shipments from outside arrive at one Community, and they need to be distributed to another. People need their coffee from the cartels in Mexico, right? It's cheaper for a shipment to be delivered to one Community, and then have it exported from there. Easier to negotiate prices."

A grin crept across Micah's face, and Cal could tell he was just getting warmed up.

"Communities are all about being private and secure, but in reality their borders are more porous. As long as you don't draw attention to yourself, you're good. Our bus looks like the ones all over Pines, so when it arrives, people are happy to look the other way. Most residents don't want to see outsiders. At all."

"Is what you have to offer to Pines really so valuable that they let you come and go as you please?"

Micah raised an eyebrow. He tapped his wrist, as if there were a watch on it. "Think about it, Cal. What do the Communities have to offer their citizens?"

Cal shrugged. "Jacuzzi tubs? Air-conditioning? Schools?"

Micah shook his head. "Yes, but all that shit represents one thing."

"Money."

Micah sighed, impatient. "Again, true. But money can't be depended on anymore. We've seen that, again and again."

"What then?"

Micah smiled. "Safety, Cal. What the people behind those gates want, and what they're willing to give up anything for, is safety. They want to sleep knowing that their house won't get robbed. They want to meet a friend at a wine bar without worrying that some maniac will blow himself up as they catch up over a bottle of Riesling."

That had happened in San Francisco not long after Micah's bombing at the mall: a man strapped with explosives walked into a restaurant in the Financial District, and ten seconds later everyone was dead. A week after that one, Palo Alto's Community announced it would be adding a new luxury neighborhood that would offer twenty-four-hour patrolling guards. A Community in Marin took another tactic and capped its membership at three thousand. The bigger the population, they argued, the harder it was to vet its members and remain safe. New members would pay a premium to live in a small and secure village with like-minded neighbors.

"Was the Group behind the other bombings?" Cal asked.

323

"The one in San Francisco was orchestrated by an allied organization," Micah said. "The others, I was led to believe, were copycats."

"What about now? What have you been led to believe now?"

"The Community that's safe is successful," Micah said. "That's their value. The safest Community can raise its membership fees and dictate prices when trading with other Communities. Once or twice, a rich Community has bought out another, less successful one."

"And you provide safety?"

Micah smiled.

"Do you, Micah?"

"I have my eyes and ears open—you already know that. I tell the guys at Pines who's out there and what they need to worry about. I report that there's a young healthy couple from L.A. living in the woods, but they're harmless. Or that there's a man and his teenage wife ten miles from here, that they're weird but too stupid to be a threat. The Land provides a travel barrier between Pines and the small outcropping of settlements to the south." He paused. "Since we got here, the Pirates have ceased wreaking havoc in the area, and they no longer skulk around the edges of Pines. That was bad for business."

"You've cleaned up the wilderness? Just by being here?"

Micah took the comic book out of Cal's hands and placed it back on the pile. "I've done stuff that would make your stomach turn."

Cal didn't doubt it. He waited for Micah to speak again.

"We cleaned up the Pirate problem."

"How?"

"We trained the boys we got from Plank. We needed to get this area under control. It would surprise you what Plankers can

324

do. Sailor, for instance? He looks like a teenage girl, but if some sack of shit's been raping and pillaging, he'll kill that fucker without hesitation."

"So you obliterated the Pirates? For good?"

"I wouldn't use that phrase, 'for good,'" Micah said. "Nothing's permanent anymore, is it? The boys and I utilized a combination of force and negotiation to solve the Pirate problem. Now they stay away from Pines, and they leave us be. Most of the people who were here on the Land when we arrived just want to be taken care of, and I gave them that."

"It must be strange working with Pines," Cal said. "How do you stomach it, after all you worked for in L.A.?"

"It's complicated," Micah replied.

"Is it?"

"Do you realize that the Communities benefited from the violence the Group enacted in L.A.? They could show their citizens, and their potential citizens, that they could protect them from all that."

"They were safe from maniacs like you," Cal said.

"Exactly," Micah said, but the calculating look in his eyes had been replaced with anger.

"We had contacts in Calabasas and in Laguna Niguel who kept the right people aware of our intentions. They knew what I was about to do."

"And you? Did you know the whole time that the Group was doing all the dirty work for the Communities? That the Group is the shill?"

"Or I was the shill." Micah shook his head, and from the expression on his face it looked like he'd eaten something rotten. "From early on, the Communities were pouring money into our organization. I had no idea."

"Did the Communities start the Group?"

"No way. I can't believe that. But somewhere along the way, they got entangled."

"'*They* got entangled'? Not *we?*"

"You keep asking if we're part of the Group."

"Are you?"

"We'd like our contacts to think so. As far as they know, we're doing what we're asked to make Pines safer, and in return the Group continues to be funded. The encampment is growing, and so is the Group."

"Isn't that what you wanted?"

"That wasn't the heart of it, Cal. Never was. I wouldn't be surprised if our encampments were eventually turned into Communities. The Group isn't what it used to be. I still believe the

Communities should be destroyed. Safety is a right for everyone."

Cal felt himself nodding. He agreed, and so would Frida.

"Toni is one of our main contacts in Pines."

"Toni? Has she been there the whole time? I thought she left the Group a long time ago."

Micah shook his head. "She isn't really in the Group. Though she is to those who need to believe it. She makes sure our permits are renewed and that our information reaches the right people. She cultivates relationships for us. And she can slip something to us into, say, Frida's beloved baking box."

"Crate. Baking crate."

Cal was picturing Toni. How she used to argue with Micah, yell at him, give him the finger, in front of other people, her posture stiff as a war general's. Toni would do anything, if it meant enough to her.

"I still love Toni," Micah said.

"I thought she couldn't hold your attention."

"I was a kid then," he said. "Now . . . I get it."

Cal raised an eyebrow.

"The point is," Micah said, "someday, we might turn on Pines. The Group wants to be their little bitch, sure, but that's not all of us. I'm forming my own Group, don't you see? Let the old members get complacent. Everyone does, eventually."

Cal blushed. He was slumped against the couch as if there were a ball game on. He'd been sitting like this ever since Micah had taken the comic book from him and had been fighting the urge to stand up again and pull one of the books from the shelves. He had his eyes on a slim blue volume of Kant. He remembered it from the Plank reading room.

Now he sat up straight.

"What are you going to do?"

"You mean what are *we* going to do," Micah said.

Bombs. He wanted to plant bombs, simple ones, ones that had the capacity to maim anyone within fifty feet of their explosions. They would kill anyone closer. He'd been building them for a few months. "That's when Sailor and Dave were brought into the plan," he said. "They have experience with explosives from their Group training, and I guess Dave was somewhat of a pyro at Plank, worse than I was. And I knew they'd follow me. They still believe in what's right."

Getting the stuff into Pines would be risky, Micah admitted. "But Toni's not fucking around. She wouldn't do anything stupid to get us caught. And there's a guy with her now. I corresponded with him back when we were students at Plank, actually. Toni said he could get us in places." It took a moment for Cal to understand that the "us" meant the bombs, and Micah meant to plant them. "Not only at company headquarters," Micah said, "and the head citizen office, that kind of thing. But also less-rarefied locations. There's a market about a mile from the checkpoint we use." Micah shook his head. "They hardly eat real food there, did you know? It's mostly weird energy bars, supplements, powdery drinks. It's why our lettuce is so beloved there." He smiled. "At their markets, they've got this spot, where people pick up their carts. It'd be perfect."

Cal didn't answer. What Micah was saying scared him, but it also sounded silly, a boy's plan, straight out of a comic book. And yet, boys were capable of terrible things.

It would take a long time, Micah continued, to get the plan in place. And once the bombs were built and delivered to Pines, they could not be set off immediately, either. Being careful was a kind of grace, he said. "We need to hurt them, shake them up."

"You want them to feel unsafe?" Cal asked.

"Not just feel it—be it."

"Micah. Listen to yourself. If Pines is rendered unsafe, wouldn't that kill your trade agreements? Won't they get more paranoid and further secure their borders?"

"We're counting on that," he said. "That's when those of us inside will act."

"It sounds like Toni will have already played her role."

Micah shook his head. "You know there's a whole underclass at Pines? Not just Pines, but most Communities. Someone has to scrape gum off the park benches, and it sure as hell isn't going to be some executive's son. Most workers have been raised within the Community, bred to do service at a place called C.A.P., Center at Pines. The kids who grow up there are called Hats, sometimes Hatters. The Communities think it's safer than letting in temporary laborers from the outside who might bring in disease and troubles of their own. The Hatters live okay, but they don't have rights, not really. They're safe as long as they shut their mouths and do their work."

"Sounds like a coup waiting to happen."

"Maybe, but you wouldn't believe how happy most of the Hatters are. They might sort through garbage for a living, but they're eating regularly, they've got access to health care, and they're given a small apartment with electricity and running water. It's more than you can say for life here." He smiled. "But not everyone there will remain content."

"Are you talking about the kids you gave up? You sent them to Pines to be part of an uprising?"

Micah held up his hand, as if Cal weren't listening, or not properly. "One of the kids, Randy, will be sixteen this year. He's bright, and strong."

"What's he going to do for you?"

"For now, he's just listening to conversations, gathering information. He's training as a security guard, so he has access. He reminds me of August in some ways. I think he can lead his peers."

"This is a long game, then, I take it."

"Toni keeps reminding me to be patient. If we do it right, Pines will fall apart: bombings, civil unrest, the whole nine yards. And once there's utter chaos behind its walls, Pines will be more vulnerable to attacks from the outside. That's a bonus. The Pirates will have a ball terrorizing those borders again."

"I thought you got rid of the Pirates."

"I told you," Micah said, "nothing is permanent. Once the bombings happen and the Pirate threat returns, those rich bastards will really be in trouble." He smiled. "It'll be just like any old American city."

"But if that happens," Cal said, "everything around here will fall apart, too. You need Pines, Micah."

"Not forever, I don't," he said. "You think I want to stay on the Land until I die? We do this, and the Group realizes I won't be put out to pasture."

"I thought you were in charge."

He nearly snarled. "I was inside the inner circle—or I thought I was. When the suicide-bombing plan was devised, I saw it as an important step in the Group's mission to change the status quo."

"And because you wanted to be deified."

He raised an eyebrow. He shook his head. "I wanted to inspire our members and make people afraid of what we could do. And I wanted to be independent, see what needed to be done outside of the Group's reach."

"But the other leaders wanted you out," Cal said.

"I didn't see it until it was too late, when I was quote-unquote 'dead,' and the relationship between the Communities and the Group became clear to me."

Micah's posture was so straight, he looked like a kid about to get measured at the doctor's. He seemed pathetic, and that shocked Cal. This was a first.

"Don't pity me," Micah said suddenly, as if he could read it on Cal's face. "Only a select few were in on my little stunt and know I'm alive. Most people in the Group are still committed to what matters, as I am." He smiled. "We'll take back the cause. We can go back to

L.A. and reclaim the Echo Park encampment. Or we can settle somewhere else until we're able to infiltrate another Community. Being a colonist is surprisingly easy, let me tell you."

"And Peter's in on this plan of yours?"

"Peter's from the Land, so his loyalties are muddled. But he's starting to get it. Especially when August, Sailor, and Dave are on my side, too."

"What about me and Frida? What about our baby?"

Micah sighed, his head in his hands. "I need you for debate, California."

"This isn't the academic decathlon. I don't want to talk you through your terrorist fantasies. Which probably won't work, and if they do, well, then, we're fucked. My kid is fucked."

"So what do you want?"

"I want to stay here. With my family. I'll gladly be a shill."

Micah said nothing, and Cal noted there was no sign of surprise on his face. Of course he knew what Cal wanted.

"Promise me," Cal said.

"Promise you what?"

"That my child will be okay. That you'll protect him and that you'll look out for your sister." Micah said nothing at first, and then, "In some ways, Frida is all I have left." Cal waited. He needed more.

"You know I'll keep her safe," Micah said finally.

It was enough of a promise for Cal. For now.

Cal nodded to the bookcase. "I want to take the Kant to my bedroom. I'll smuggle it under my shirt if I have to."

Micah laughed in his face. "Hell, no."

Cal laughed, too. He could picture the title page, a mimeograph of the original. He could smell the book's interior: like almonds and wood chips, the glue sweet as warm milk. He closed his eyes for a moment and imagined that scent. And then he thought of the Forms in the dark, how he'd understood them, how he had anticipated each one before he passed it, as if he'd known them all his life. He would need to keep himself here. He would need to help Micah, but not the way Micah wanted him to.

"Let's figure out another plan, okay?" Cal said.

He wouldn't tell Frida any of this, at least not until the plan had been perfected, and maybe not even then, not if keeping it a secret meant she would sleep soundly at night. She needed to rest for the baby. She would be happier not knowing, as long as he had her best interests in mind. As long as he kept demanding information from Micah and was being smart, she'd be satisfied. She could trust him to make decisions for their family.

Cal sat down on the couch again. "There's got to be another way to return the Group to its pure beginning, to cause problems for the Communities." Even as he said these words, their cheesy call to arms, their rah-rah-rah cheerleading, he felt their power. He did want to find a better way.

"I'm listening," Micah said.

"I have no idea what that is yet. But there has to be something. There's always another way to approach the text, isn't there?"

"Oh, baby, talk nerdy to me," Micah replied, but he was listening.

"There has to be a better plan," Cal said.

Before they left the room, Micah went to the bookcase and pulled out the Kant.

"Stuff it into your jacket, and don't let anyone see it. And I mean it, Cal, not anyone, not even Frida. If you do, I swear I will cut off your balls with a paring knife."

Cal took the book, nodding. It was his victory, and both of them knew it.

19

Frida couldn't tell if she'd overslept because it was always dark these days when she woke up. She'd fallen asleep to the sound of rain, imagining the Land turning soggy and slippery as she remained safe and dry inside the Hotel, but all was quiet now. It must have stopped. Good. Cal had spent the last few nights on security, and Frida didn't want him getting soaked and sick.

Now that the boards had been nailed to their bedroom window, it was night all the time. The darkness and damp and the smell of people sleeping reminded Frida of the Millers' house. On the coldest days, she and Cal used to crawl into bed, into that corner where the mattress fit perfectly, and force themselves to sleep as long as they could.

"We're hibernating," Cal would say, and reach for her.

She'd been so bored with that one-room house and the woods surrounding it. That grimy outdoor cooking pit of theirs, it would never get hot enough until it got too hot, and that same door to look at when she woke every morning. Sometimes even the sound of Cal's voice, his stiff walk, how he held his mouth when he was being serious, had bugged her. She'd been so sick of their isolation. And now look at her, she was imagining that old life with something bordering on longing. Dada had always called her capricious. Maybe this was what he was talking about.

The first time they were alone after they argued in the Bath, he'd said, "I'm doing what you asked." He had pulled her to him, and kissed her.

Whatever he meant by that, Frida felt comforted. She wanted it to be enough. It had to be. Cal was offering her the only solace available, and she took it because it helped push the gruesome images of her brother out of her mind: Micah using a large knife to behead the Pirate; threatening Anika with that bandanna; taking the Bee from Ogden. Did the baby wail out for the toy, refusing to let go, or was he asleep, and Micah nimble as a thief so as not to wake him?

"I found out what happened to the children," Frida had said.

"So did I."

"Micah told you?"

He nodded. "We have to remember that not everyone on the Land had children. And those who did knew they were giving their kids a better life. It wasn't cruel, Frida. You see that, right?"

"What about the older children?" Frida asked. "They weren't adopted. Did you find out about that?"

Cal didn't say anything.

"Cal?"

"He won't touch our baby," he whispered. "Micah needs us here. He won't let us be exiled."

"You really think Micah will protect us?"

"You're his sister. And he needs my help."

"But don't you think some people will be upset about the pregnancy?"

"I don't know, Frida. We need to wait and keep watching. With a little more time, I think we can win them over. Micah will make them see that it's for the best. He's good at that."

"That's true," she said.

Frida let him kiss her again. He'd said he was doing what she'd asked, and she decided that meant he was looking out for her. Since their fight, he'd been attentive and gentle, actively seeking her out after Morning Labor, seeing if she needed anything. He was paying attention to her again. He hadn't gotten lost in the dark.

Frida got out of bed and got ready to head downstairs to the kitchen. She was just pulling on a sweatshirt when Cal entered with a flashlight. He was wearing a raincoat, but it looked dry.

"You're still in here," he said, surprised. The flashlight's beam bounced across her and then paused on the unlit candle by their bed. "I didn't see any light coming from under the door, so I assumed you'd left for the kitchen already."

"I can get dressed without a candle," she whispered.

He kissed her and put down the flashlight so that its light spread across the ceiling.

"But why?" he said, heading to the candle. "Let's splurge."

The flame flickered and rose, and Cal turned off the flashlight.

"I'm going to the campfire tonight to talk to Anika," she said.

She'd decided that she would be up front with Cal, show him she could gather information, too.

"I'm going to hang out here," he said, "if that's okay."

"Of course it is," she said. "I think I'll have a better chance of talking to her there. I haven't been able to since Fatima started baking with us."

"I thought Anika already told you everything."

"She did."

"You haven't said anything about the baby, have you?"

"You know I haven't," she said.

"I know."

"I guess I still feel unsettled," she said. "Like, I need to see that this place is good, despite all that's happened."

"It is," Cal said. "It will be."

He was sitting on the bed now, and she stood before him. She placed her hands on his shoulders, her eyes on the far wall. She imagined herself on the deck of a majestic ship. She would just have to keep reminding Cal that they'd come here together, and that, if necessary, they'd leave that way, too.

"I love being married to you," she said.

Cal smiled. "I could live off those words," he said, and pulled her toward him.

———

There were at least fifteen people sitting around the campfire, talking loudly over one another as if drunk, passing a cup of something hot, poured from an ancient metal thermos. Frida thought she could smell mint tea, but that had to be her imagination because the air was so smoky she'd started breathing through her mouth. The scene reminded her of the beach, cooking oysters in sand pits with her parents on a trip to Northern California when she was thirteen, before they'd had to sell their second car. She wasn't sure what she'd been expecting of this campfire, but it wasn't this. This was a party.

Anika looked up at Frida when she arrived but didn't wave her over or even nod. Peter was there, too, plucking at a guitar in a rickety lawn chair, trying to recall a song. He hadn't seen her. Frida tucked herself behind Anika on the shower curtain where she was sitting. The perfect waterproof picnic blanket.

Anika hadn't told Frida anything since Fatima had joined them in the kitchen; no doubt, that had been Fatima's goal.

336

Someone had asked her to intrude on their privacy, and she had complied. How many times had Frida wanted to tell Anika she was pregnant? Just to lean away from Fatima and whisper the news. Anika might be upset at first, but not when the reality of Frida's pregnancy settled in. There would be a child on the Land again. Anika could be Frida's guide. She could be the child's aunt.

Frida had so much to say to her. She wanted to know about Ogden's birth, for one, and ask her about diapers and clothing. She wanted to tell Anika that she was certain she was having a girl. A daughter.

Frida looked around and realized with the noise from the fire and singing and guitar, they could talk fairly openly without being overheard.

"Fatima's in the way," Frida whispered finally.

"Fatima's a bitch," Anika said. "She came a few weeks after Micah, you know, with the rest of his settlers. She was real close with August."

"Were they a couple?"

"They claimed to be just friends. Not long after she arrived, she became Peter's girl."

There was reproach in her voice, and Frida realized that Anika really did hate Fatima. For taking Peter. For simply having a partner. Or for treating herself as chattel, passed from one man to the next. Or for joining their morning sessions without asking first, for babysitting them.

Babysitting. The baby. It always came back to that. She had to keep it a secret until Micah thought it was the right time. Definitely not before the Vote.

Maybe Cal was right: he and Frida could help make the Land into the place they needed it to be. She wanted to ask Anika if such a dream was possible.

Betty came over and sat down next to Frida, who had scooted over to make room on the shower curtain.

Betty rubbed her hands together, her face to the sky.

"Cassiopeia," she said, to no one in particular.

Frida looked up but saw only white clouds.

"In theory," Betty said, and laughed.

Lupe and Sheryl came over and sat next to Anika. Frida liked Lupe, but Sheryl—what had Cal called her?

A stick-in-the-mud. That was a nice way of putting it.

But now, looking at their backs, Lupe's slumped, Sheryl's straight, Frida saw a closeness between the two women that she'd never caught on to before, and it made her happy. It was the casual intimacy of old friends; they had shared beds, swapped shoes, probably undressed in front of each other dozens of times, kept talking as one of them peed. If they had looked anything alike, they might be mistaken for sisters.

Frida watched as Anika, without speaking, passed first the cup and then the thermos to Lupe, their fingers briefly touching, and she realized all three women must have started the Land together. With Sandy Miller, too. They probably had known Jane. They remembered Ogden; maybe they had advised Anika on what to do. Or they had given away children, too. They probably still avoided red and certain stories. They had accepted Micah and his way of doing business. Maybe Anika didn't trust Micah, but Lupe and Sheryl probably believed things were better with him around. Maybe Sheryl wasn't that bad; maybe she was just prickly like Anika.

Peter had finally found the correct chords for the song he wanted to play, and the other side of the circle began singing along with him. Frida couldn't place the song, though she thought it was a ballad from the last century, something her father might sing as he made dinner, humming everything but the chorus. As the voices rose, earnest and off-key, Betty leaned forward and whispered to the women in front of them, "I hear Rachel's sleeping with Dave."

Frida could tell the women had heard Betty by the way they looked at each other. Sheryl had the cup and thermos now, and she snapped them together so forcefully that Lupe laughed.

"Settle down, Miss Sensitive," Lupe said. "It's not as if he's any good."

If Frida had been drinking anything, she would've choked on it.

"Dave?" Frida said, without thinking.

This time, Sheryl turned around. "Anika said you were cool."

"She is," Anika said, still facing the fire.

Betty put a hand on Frida's knee. "We're warm-blooded creatures."

"But Dave is so young," Frida said. These women were old enough to be his mother, but she didn't say that. She knew she sounded prudish already, and she didn't want to be nudged out of this locker room too quickly. "I mean . . . good for Rachel."

Lupe laughed, turning around. "Sheryl, give her some of the milk."

Sheryl unscrewed the thermos slowly. "She doesn't look thirsty to me."

"Oh, but I am," Frida said.

It was cow's milk, heated to a foam. It smelled oddly sweet, like the postage stamps her grandfather had collected for nearly his whole life. He liked extinct things. He'd given Frida one for her eighth birthday, and she'd licked it as soon as she was alone in her room.

"So there aren't rules against it?" Frida asked Betty.

"Not officially." She smiled. "But you must have seen the whole drawer of Pills in the Bath."

Frida was surprised, but she knew she shouldn't be. Micah didn't want children here, and he'd make sure no accidents happened.

Likely, the women were grateful to have access to birth control, certainly procured from Pines.

"Otherwise, when it comes to love, we can do what we like," Betty said. "As long as we're discreet, that is." She grinned. "And those who prefer to abstain pretend everyone prefers that." She nodded at Anika.

"I never said that!" Anika said. "It's just my own personal choice."

"You're our nun," Betty said.

"Ha," Sheryl said. "You and Micah."

"My brother?" Frida said. "Really? He used to be such a dog."

"He's too serious for all that now, I suppose," Anika said.

"The fact that he's never been interested in sex made us like him," Betty said. "He never touched us."

The women fell silent.

"Who knows what happens on his treks off the Land," Lupe said.

"Micah leaves?" Frida said. "With August?"

"Not often," Betty answered. "Occasionally he needs to help August at Pines. They have to lug these big containers of soil. Sometimes cantaloupes or lettuce, whatever it is that we're trading that month."

"I bet he just wants to get away from us," Lupe said. "It can be pretty boring around here. Maybe he has a secret wife living in the woods."

The other women laughed.

"Micah? Can you imagine him having a wife?" Sheryl said as she reached her hand out behind her. She wanted the cup back.

Anika turned to grab the cup from Frida. "I hope you weren't hoping for a top-secret meeting of the minds tonight. It's just us ladies, gabbing."

"Sure, it is," Frida said.

"Micah should bring his secret wife here. It's not like someone is going to try to steal her," Lupe said. "No one goes after other people's partners here. Monogamy is respected, rare as it is."

"She's saying you don't have to worry about your husband," Betty said.

Frida shook her head. "It didn't even occur to me that I should."

Now all four women were looking at her closely, as if trying to gauge her truthfulness.

"Not even his eye wanders," Sheryl said, oraclelike.

"I don't know about that," Frida said. "But he's a good man."

"You don't worry when he leaves you at night?" Anika asked.

"He's with some of the other men," Frida said. She paused. "And at dawn he goes to see my brother. And, Lord, if they're being amorous, I don't want to know."

The women laughed, almost loudly enough to be heard over the singing voices. The song sounded melancholy now, the key too high for most, but still the singers tried to reach it.

"The Vote's coming up," Anika said. She was looking at Sheryl.

"I know," Sheryl said. She turned to Frida. "Glad you could make it here tonight."

So Anika wanted Frida to get to know her friends, to prove to them that the new girl was cool. That she could fit in. That she was worth voting for.

"I'm glad I could come," she answered. She smiled. "If you keep me on the Land, I'll let you sleep with my husband." She winked at Anika.

Sheryl snorted.

"I thought you'd never offer!" Betty said, and they all laughed again. Frida felt the pride that being funny brought, and for the first time in days, she felt happy, and safe.

The women left before the fire got too low, but Frida stayed. She wanted to listen to the singers for a little longer, she said. She promised Betty she'd return the shower curtain picnic blanket to the outdoor lounge. The night air, scraped clean by the rain, felt good on her face.

Rachel showed up soon after the women left. Without the guitar to guide her, Rachel began to sing "This Little Light of Mine."

Her voice was deep and scratchy. She had been a smoker in her past, Frida could tell. She imagined Rachel twenty years earlier, Dave's age: her hair long, lots of eye shadow maybe, definitely a run in her tights, drinking a lot, every night her lips stained purple with wine. Back then she wouldn't have sung, unbidden, like she did now. Dave probably wouldn't have liked her young. Hilda had once said that some women, the lucky ones, lost their youth but found something much better, something sexier, to replace it.

Frida was still young, though, wasn't she? She was sort of in between the younger men and the older women. That couldn't have escaped notice. The women must wonder if she wanted children, now or ever. They must have considered her fertility. If they felt threatened by her youth, they didn't show it.

The next song was one she knew, and Frida decided, what the hell, she would sing along. She remembered the lyrics from day camp so long ago, when there was still money for that kind of thing. She wondered whether Micah had supplied it to the Land's canon.

Frida sang loudly and terribly, and she laughed with everyone else when Peter supplied the baritone echoes. She wanted this. She wanted to stay here. It was what she'd wanted when they arrived, when she had fallen into her brother's arms, and when she met all these strangers, saw their buildings, ate their food. She had wanted to be part of a community, and, abracadabra, here it was. She'd felt so lucky. That feeling was coming back to her now.

Once the fire died down and everyone began to disperse for bed, Frida's secret surfaced in her thoughts once more, and she wasn't sure what to feel. She was a fraud. She was a liar. Her friends were all following the rules of this place without complaint, and now here she was, an exception to those rules. It wasn't right. If Frida thought Anika would be happy, she was crazy. Anika would come around to the idea, Frida was sure of it, but the longer the pregnancy was a secret, the worse it would be. Frida would look no better than Micah, who fed off secrets. She and Cal would be starting off here badly if they withheld this information.

She tried not to think about it. She wanted to push the baby from her mind. Not now, not now.

Already sounding like a mother, she thought.

When Peter caught up with her on the way back to the Hotel, there was no use wasting time with a preamble. "I feel like people should know," she said.

"They will soon enough," he said.

He seemed so glad that she was pregnant, despite the complications that were sure to take over this place, at least for a little while.

Frida smiled. "It's wonderful, isn't it?" She couldn't help it.

She was channeling Sandy Miller, she realized, triumphant

before her chart of menstrual cycles, glory be the gift of children, excited for the bounty they would inherit. Because that's what moms did, right? They chose to believe the future was good. To assume otherwise was to participate in a kind of despair.

Peter squeezed her shoulder and told her to sleep well. "August will be back tomorrow," he said, and she nodded. The Vote was upon them.

And here she was, a few nights later, sitting in the Church next to Cal, in the same pew they were always led to, right up front so that nearly everyone was behind them. The first time, she'd been too shocked to really take anything in; her brother was alive and here was a whole town of people just a two-day journey away from the Miller Estate. The first night, she couldn't possibly have been bothered to notice that someone had carved the initials d.b.b. into the pew's wooden seat or that the buzzing lights at the back of the room seemed to be saying *uh-huh, uh-huh* over and over again. Cal had initially found those lights to be obnoxious, but he didn't seem to mind them now. He had nodded at a few people on their way in but had since fallen quiet. He squeezed Frida's hand every now and again, and she squeezed his back.

Betty had told her that housekeeping never cleaned the Church's interior, but clearly someone had been in here to dust. The stage before them was clean and buffed, the metal piping around its edge smooth as the hem of a gown. There was nothing on the pulpit: no ballot box, no table with small slips of paper, no vat of ink to dip people's thumbs into after they'd cast their vote.

Then Frida remembered that it would all happen publicly. That's all she knew.

She hadn't asked Micah or Anika, and it seemed odd to turn around and ask Rachel, who was sitting behind them. Dave was sitting elsewhere, of course. Rule 1: discretion.

"Do you know how it's going to work?" Frida whispered to Cal.

"Everyone who wants us to stay will move to a designated corner of the Church," Cal said. "Anyone who doesn't will go stand on the opposite side."

Someone whistled, a piercing, two-fingered one, and Cal stopped talking. Micah walked onto the stage.

"Let's get started," he said, and clapped his hands twice.

Frida had expected her brother to say a few words about the Land's philosophy, about how this was a significant moment in their history as a community. They hadn't accepted any new members since he and the others had arrived a few years back, and that had to be on everyone's minds. It wasn't until Micah didn't say any of this that Frida realized she'd been composing a speech for her brother in her head these last couple of days. She had imagined him describing Cal's gifts as a farmer and carpenter and critical thinker. He would go on and on about Frida's bread, about how well liked she was. He might say something about family. *Frida is my sister,* he would say, and leave it at that because everyone would understand how meaningful that was.

Instead he held up his hand and said, "I'm confident that everyone has already made up their minds." He paused, and Frida imagined everyone behind her nodding. "So here we go. If you're in favor of Frida and Calvin moving to the Land permanently, to participate in our community, please move to the northeastern corner of the room." He pointed to the back-left

345

corner of the Church. To Frida and Cal he said, "Please remain seated, guys." And then he jumped off the stage, presumably to make his own vote.

"Don't they want to debate it?" Frida whispered to Cal. "They don't have questions or anything?" She felt so ignorant. It hadn't occurred to her to ask these questions earlier.

"A few took private meetings with Micah and Peter," Cal said, "to voice their concerns." His eyes remained on the stage, but Frida could tell he was listening closely, trying to discern the migration patterns of those in the pews behind them. "August's been lobbying for us the past two days."

Cal was obviously confident they had everyone's support, and after a few moments Frida felt him relax against her into the pew.

"You're acting as if there's a screen in front of us," she said. "Like we're at the movies."

He smiled. "Pass the popcorn."

He must have pushed the baby out of his mind. That, or he truly believed that once they were accepted here they could not be forced out.

From the corner of her eye, Frida could see that the people in the pew to their left were moving across the room. She turned her head, expecting seriousness on everyone's face, but Peter was absently running his tongue over his teeth as he passed the last pews, and Fatima had gotten distracted by her thumb-nail; she almost bumped into August, who was just ahead of her. Anika was smiling, for once.

She could tell Cal didn't want to look at the gathered group until everyone had finished voting and that he expected her to do the same. She didn't care. She turned to watch the people crowd into the far back corner. They were a disheveled and unlikely bunch, huddled together as they were around the campfire. She'd been to it every night since the first time. Yesterday she'd brought cinnamon buns to pass around, and Sailor, a rare visitor to the festivities as far as she could tell, had joked, "Buttering us up for the Vote, lady?" before taking two. Everyone laughed, including Frida, but it had also made her uneasy. He was right. She had been deliberately campaigning, befriending anyone who looked her way. As if she were running for prom queen.

More and more people had clotted into the northeast corner, and after a few minutes it appeared as though everyone had moved out of their pews. Only Frida and Cal remained seated.

This was good, she told herself. She and Cal had been accepted. They were wanted.

But even as relief passed over her, so did its inverse, its shadow. It was the same shame she'd felt flush with at the campfire. She just wanted them to like her, and there was something selfish about that, especially when they didn't know about the baby. The baby was important; it was necessary information.

Tell them.

Was it crazy to imagine her baby, passing on this message? It was as if Frida had picked up a bottle that had washed onto shore. She had unfurled the scroll to find these instructions. *Tell them.*

"They need to know," Frida said to Cal.

"They will," he said. He was still facing straight ahead. She remembered what he'd said a few nights before. *Wait and watch.* He actually thought Micah would figure it out for them.

But would he?

"No," Frida said. "Now."

She felt herself standing. Cal's hand had grasped her own, he was trying to yank her back to the pew like a current pulling her underwater, but she shook him off.

"Micah," she said. Too quietly at first. The collective volume of the room had risen suddenly. Everyone had begun to talk to their neighbors; they were excited, Frida supposed, by the official change, by the obvious outcome of the Vote. The Land was growing! They could not be contained!

"Frida," Cal said. "Please."

She didn't look down at him.

"Micah," she said again. She yelled it.

This time, everyone heard her, and her brother emerged from the center of the crowd as if he'd been pushed forward.

Everyone had stopped talking.

"You okay, Frida?" Micah asked. "We usually hold an optional postvote analysis, if you want to contribute then."

"Let's wait," Cal said, but not to everyone, only to her.

Frida sought out Anika's face in the crowd, but before she found her, she saw Betty and then Lupe. And Rachel. Had all these women been mothers? Her eyes passed over Smolin. Had he been a father? If these people had been parents once, they still were. That role could never be taken away.

Her parents had grieved Micah's death; their son was dead, but they were still his parents.

"I'm pregnant," Frida said. She said it loudly, she made sure of that, but she repeated herself, just to be sure. "I'm pregnant."

The lights huffed over the silence that Frida's news had wrought. She looked immediately at Cal, who had let go of her hand. His eyes were on his lap.

"Excuse me?" Micah let out a harsh and sudden laugh that startled those nearest him; it was as if he'd punctured a balloon with a needle. "Did you just say you're *pregnant?*"

His delivery was perfect.

Peter stepped forward, with the same innocent, confused

expression as Micah's. Frida couldn't help but be impressed with his acting, too. These guys were good.

"I wanted to tell you all before," Frida began. She realized she had no excuse that wouldn't implicate Cal, who she'd promised was a good man. And she didn't want to tell on her brother; if she did, the Land might not recover.

Cal stood up. "I asked her not to," he said. He had taken her hand again. "I thought it would make it easier. I wanted you to consider just us first, before anything else."

All at once, people began to murmur to one another. Frida felt them looking at her, as if scrutinizing her body for signs, for proof of her betrayal. She wanted to lay a hand across her belly, but she didn't. That would be too much for them.

The volume kept rising. It was like the back draft of fire, enveloping them with a *whoosh*.

"I can't believe this," Micah said, somehow louder than the others, and the room fell quiet again. "Frida?"

"Oh, please," Charles called out. "You expect us to buy that, Mikey? Your sister is pregnant, and you don't know?"

"You know everything." It was Sheryl's voice, but she was behind the others, and Frida couldn't see her. "You knew. You had to."

Micah said nothing, only shook his head.

"What does this mean?" Fatima asked.

Charles nodded. "Yeah. We can't just change everything we've come to stand for. What about our rules? *Your* rules?"

All at once, people began to walk away from the corner.

"Everyone, please remain calm," Micah said.

"Where are they going?" Frida asked Cal.

"They're voting us out . . . or they're coming for us." Cal was looking at Micah and Peter, whose heads were bent toward each other, whispering. August was headed to their huddle, and Sailor and Dave stood a few feet away, alert as bodyguards.

"Or maybe the conversation will continue," Frida said. "Some of them might want it."

She meant that they might want to talk further about what the future would look like if it had a child in it. But she also meant the child itself. Her baby. Would anyone be happy for her?

She looked back at the corner for Anika. She could only imagine what Anika must be thinking. Of Ogden, maybe. Babies are newborn for such a short period. Frida wanted to tell Anika that she would get that beautiful time, if not back, then again. They all would.

But Anika wasn't standing in the corner. She'd left it. Something hooked into Frida's gut, and she turned swiftly around. There was Anika, crowding the aisle with a group of others. The room had grown loud again, and Frida yelled Anika's name. She didn't care who heard.

Anika looked up, right at Frida. Her face was flushed with anger. Her eyes were hard and black, full of resentment and rage.

Anika turned away quickly. A forsaking. "I think Sheryl's right," she called out. She was standing again. "Micah had to have known."

At that, Micah skipped onto the stage; Peter and the others had returned to the pews, except for Sailor, who stood by the door—with a gun, Frida saw with an intake of air. Cal had her arm now, and he leaned in to whisper, "I've got you."

"I assure you, Anika," Micah was saying loudly. "Had I known, there would have at least been a transparent discussion about the matter."

"What's there to discuss?" a woman called out; Frida could not bear to look to see who it was. "We believe in containment!"

A few people cawed like peacocks, and everyone started

talking, and Frida closed her eyes. She didn't want to see all of them holding up their fists in agreement, knocking the air.

"We'll just go back to the Millers' house," she whispered to Cal. "Just like before."

He shook his head, as if that wasn't an option.

"We have to," she said.

"This is why I told you to stop," he replied. He was still touching her protectively, but his voice was cold as the dead. "You fucked it up."

"You think they'll feel differently next week? Or next month?"

"Micah and I were working on a plan," he said.

Frida was considering how to answer him when Micah whistled to shut everyone up. "Things are spinning out of control," he said.

August nodded. "Let's calm down."

"I don't want my sister and her husband to be fearful," Micah said. He held up a hand.

"You think we'll hurt them?" a man called out from behind Frida.

"Yeah," Anika cried, "that's *your* specialty."

"Oh, for God's sake, Anika," Fatima said. "Micah's done a lot for the Land. For you especially. Don't pretend like he hasn't."

"And he'll do the most for them." Anika pointed at Frida and Cal.

"Those two can't just come in here and expect us to change everything for them," Sheryl said.

The volume of the room began to rise once more

"Get them out!" Lupe yelled.

Micah tried to hold up his hand again to calm everyone, but it didn't work. Frida felt the ceiling and the walls of the Church wobble and contract, as if she and Cal were about to get crushed from all sides. They had to get out of there. Sailor had that gun, but he couldn't protect them against these people, this mob.

"Cal," she said.

He was already pulling her out of the pew. Charles had tried to step in their path, but Cal pushed him hard, and he backed away. August had another man by the arm, holding him back from who knew what.

Betty made a retching sound. She was about to spit on them.

Cal pulled Frida down the aisle and toward the door. No one else tried to stop them.

On their way out, she saw that Sailor had his rifle up, aimed at the rest of the Land's residents. How long before he turned the gun on her?

"Keep going," Micah said from behind them. "Run."

20

They rushed out of the Church, and Micah took the lead, running in front of them as if Cal and Frida were chasing him. They ran down that dusty path the rain had turned to clay, across the open soggy field, past the garden, and into the woods.

Cal was trying to remain placid even as his chest burned, even as he wanted to stop and catch his breath. He knew Frida had good intentions, she only wanted to be honest, but it was hard not to just let go of her hand. She couldn't keep her mouth shut to save her life—to save the life of their child. He'd thought they were in this together. But no. God, she could be such a selfish brat.

Cal could tell Micah was nervous. It was his speed and silence that gave him away, his refusal to turn around to see if they were still behind him. It was as if he didn't care that they might get lost trying to keep up with him. For Micah, it wasn't about his sister, or the baby. Screw all that. Micah had to worry about his plan. It wouldn't work if there was a revolt.

His plan. Talk about a euphemism. In the last few days, Micah had shown Cal the materials he was hoarding to make the bomb, hidden in the Church's secret library. "Weapons and books," Micah had said, grinning, "the perfect combination." Once he got everything together, he said, August would do a test.

"We need to make sure they work," he said. "Toni is adamant about that, as she should be."

Cal had just listened and asked questions, considering everything from every possible angle. He would have to come up with a feasible alternative soon and convince Micah that his idea was shortsighted and should at least be delayed. In the meantime, he'd just let the man talk.

Now, in the dark, Micah stopped suddenly, and they almost ran into him. Cal had managed to hold on to Frida's hand, though he couldn't bear to look at her.

Micah pointed to the trunk, and Cal realized they were at the tree house. "You'll wait there until dawn," he said.

"And then what?" Frida said.

"Maybe you should've thought about that before," Micah said.

"I thought some people would be happy about the baby after what happened here."

Cal could feel her looking at him, and he turned. Even in the dark he thought he could see the line between her eyebrows deepen. She wanted him to respond.

"Up," Micah said then, saving him.

It was cold and wet in the tree house, but there were blankets up there; they'd survive. Micah had lit a large candle, though he would extinguish it when he left. "You're hiding, remember."

"What are you going to tell them?" Cal asked.

Micah put a hand above the candle's flame, as if seeking its warmth. "You're better off not knowing."

Cal leaned against the tree trunk, silent. He just wanted to close his eyes.

Suddenly, Frida's hand was on his arm. "I'm afraid for our child's life," she said. "Aren't you?" she asked.

Cal was going to say, *Yes, more than anything, yes.* He wanted to say, *That's why you should have kept quiet.* Instead he said, "It's okay, Frida, it's going to be okay."

But she was looking at her brother.

"You sent all those kids away," she said.

"I did," Micah replied, and he was so calm about it, Cal could sock him.

"It's not something to be so blasé about," Cal said, a panic rising in him. Micah had said he'd keep Frida safe, but did that mean he'd protect the child, too? "What about our baby?" he asked.

"What about it?"

"Would you have taken my child away from me?" Frida asked.

"There is no *taking*," Micah said. "I never took anything from anyone." He gestured to the trees around them, some still dripping with water from all the rain. "You really think you're special? That just because you're family, you mean more to me than anyone else?" He sighed.

"Is that what the Group taught you?" Frida said.

Micah grunted. "Hardly. The Group taught me that some things are worth sacrificing everything for." He paused. "And then they taught me not to trust anyone."

Frida put her hand on her belly, as if to shield it from a blow. "Was it really necessary to break up the families?" she asked. "Or did Pines make you an offer for those kids that you couldn't refuse?"

"This isn't about my powers of refusal."

"Anika loved Ogden."

"Ogden?" Cal said, and then shut up. Of course—Anika had been a mother.

"Ogden was still an infant when he went to Pines," Micah

said, "and he was sick a lot. He was underweight. He couldn't hold his head up." He rubbed an index finger along the wood planks between them, as if to erase a smudge. "I was doing them a favor. I saved his life."

Frida didn't answer, and Micah said, "You have no response because you know it's true."

"What about Jane?" she asked. "Did you think she'd die out in the wilderness?"

"To be honest, I wasn't sure," Micah said. "No one was here against their will, not even Anika, no matter what she tells you." He paused. "The Millers had more skills than anyone else on the Land, which was partly why they were eager to live on their own. They were very independent. I admired them, actually."

"Why didn't Peter and Anika go with them?" Frida asked.

"Peter and Anika?" Cal asked. "Why would Peter go any-where with Anika—"

"Catch up, Cal, will you?" Micah said.

"Answer my question," Frida said.

Micah smiled. "I didn't encourage it. I told them leaving the Land, even together, was asking for trouble. Just look at how they fared before I'd arrived. And, anyway, Peter was happy Ogden could go to Pines —he'd be safe there. And he wasn't with Anika by then anyway. The point is he wanted to stay on the Land, and I was grateful. I offered to help the Millers get settled, to send a team to build their house, and to make sure August came regularly."

"In exchange for what?" Cal asked.

"That they stay away from us, live far enough away that we would never see them again." He turned to Frida. "Anika didn't tell you about this little negotiation?"

Frida shook her head.

"Not that the Millers ever expressly invited Anika to join them. Sandy always thought she was a bit uptight."

"Don't be cruel," Frida said.

"I'm being honest," Micah said. "The truth is, it doesn't matter what anyone thinks of Anika. Her relationship with Peter gives her a modicum of power. It's over between them, of course, has been for a while, even before we arrived. But those two have history. Peter looks out for her." He grimaced. "She has to be tended to."

"Peter wants me to have my baby here," Frida said. "He's excited, I can tell."

For the first time since they'd ascended the tree house, Micah looked rattled. He stood and grabbed the railing, as if to anchor himself from a strong wind.

"Peter's just being friendly," he said.

"You sure about that?" Cal asked.

Micah nodded. He was already backing away; he was already trying to leave them here.

"Micah," Frida said.

"What is it, my dear?" he asked, his voice icy.

"Why didn't you let Ogden keep the Bee?"

Cal looked at his wife. What the hell was she talking about? "What's the Bee?" he asked.

Micah wouldn't even look at Cal. His eyes were on his older sister.

"I thought there might come a time that I'd need it again."

"For what?" Frida whispered.

"Imagine what would happen if we had another baby here. There's protocol: we believe in containment, and children aren't allowed."

"What's the Bee?" Cal asked again, but neither of them was listening to him.

"Think about it, Frida," Micah said. "I'd have to keep the little thing happy on the bus, wouldn't I?" He paused. "The ride to Pines is long for an infant."

Frida stepped back. "You would never take—"

"Not yours, I wouldn't."

Micah knelt down and blew out the candle, leaving them in darkness. "I've got to go," he said, and in seconds he was gone.

21

Frida woke to the tree shaking. Someone was climbing the trunk.

She grabbed for Cal, who was curled against her like a cat. Even in his tight embrace, she was cold.

"Cal," she whispered. "Someone's coming."

Frida could only make out the edges of the platform and the shape of Cal's body against hers.

She was sitting up when Anika hefted herself onto the platform, a small flashlight between her front teeth like a dog carries its bone. When she saw Frida, she took the light out of her mouth and shined it into Frida's eyes.

"Anika," Frida said, pushing herself to sitting.

Anika didn't speak. A shapeless Batik purse was slung across her body, bisecting her chest with its strap.

In one deft move Anika had replaced the flashlight with something else. The moonlight wasn't much, but Frida recognized the size and shape of Cal's gun. Frida had held it in her own hands in the dark many times.

Why had Micah left them unarmed?

"Hey," Cal said, and untangled himself from Frida. When he saw Anika he said, "Be careful with that."

"What's happening?" Frida said.

"You don't recognize your own gun?" Anika said to her. "Or

let me guess, it's your husband's." She cradled the weapon in two hands now, as if it were fragile. At least she wasn't reckless enough to point it at them.

"Where'd you find that?" he asked.

"Hidden in your room."

"You went into our room?" Frida said. "Micah said Dave would be guarding the door."

Anika laughed. "He's supposed to be, but Rachel helped us out with that problem. When he thinks with his dick, he's an easy target, poor thing."

"Rachel?" Cal asked.

Anika nodded and smiled at him, a dark space where her tooth should have been. "Don't be shocked," she said. "Rachel is quite alluring to the younger set."

Cal stood up, and Anika took a step back.

"Stay where you are," she said. With one hand she reached into her purse. "I found the gun under your bed. It was hidden with your book."

"What book?" Frida said.

Anika held up a hardcover book, its spine duct-taped. Frida hadn't seen one of those in years.

"What is that?" Frida asked. "Whose is it?"

"My question exactly," Anika said. "Now at least I know you didn't arrive with it."

Cal shook his head. "Micah has a few books. He doesn't share them with the rest of you, and I agree it's wrong. I told him."

Anika tossed the book at him. He held up his hands to block his face, and Frida couldn't help but think he looked like a pussy.

The book fell to the floor, its covers splayed like a bird's wings.

"I don't give a shit about the book," Anika said. "But I know others might. It's just like Micah to hoard it."

"My brother shouldn't have kept it from all of you," Frida said.

Anika groaned. "Oh, please, Frida. Like you're some expert on ethics."

"I wanted to tell you about the baby, I swear I was going to."

Anika had the gun in one hand now, and it looked like she was going to cock it.

"Let's put the gun down," Cal said. He kept his voice low and gentle, his body still. It wasn't working.

"I told you about Ogden," Anika said, "about losing him. And all this time you were pregnant, expecting to have your own baby. Things must always work out for you, Frida. You're always the exception to the rule."

"That isn't true," Frida said.

"You betrayed me."

"Put the gun down," Cal said. "We can talk this through."

Frida wanted Cal to just tackle Anika; he could do it easily.

"I'm sorry I kept this from you," Frida said. Would Anika really fire the gun? "But think about it," Frida said. "You could help me, help us."

"What makes you think I want to help you?" Anika asked. "Why do you get to have whatever you want? Because he's your brother?" She laughed. "You know what he did, right? He came back to the Church tonight, said he'd taken care of things."

Frida nodded. For a long time, she'd lain in this tree house thinking of Micah and his threats. He claimed he wouldn't send their child away, but was that enough? He'd let this place come between her and Cal. He'd willed it, even. Her brother was cruel, sometimes in such small ways it was barely recognizable. He loved like that, too.

"He told us we could go back to normal," Anika said. "People think you guys are either dead or that Mikey sent you into the Forms without even water. But I didn't believe it."

"Anika, put down the gun." Cal stepped forward. "Frida and I can leave the Land tomorrow morning if that's what you want."

Anika turned to him. "Okay, go. But don't expect us to help you like we did the Millers. We were protecting them. We cared about them."

"Why didn't you take Ogden somewhere on your own?" Frida asked. "What kept you here?"

"Frida," Cal said. "Stop."

Anika didn't speak for a moment, just kept cradling the gun. She looked up. "Do you have any idea how hard it is to raise a child out here? Before Micah and the others arrived, we were barely living. It took everything I had to nurse Ogden. And after that he still had to be kept warm, and entertained, and I had to sing to him when he couldn't sleep, and I had to make sure the water I drank wouldn't kill me, or that a Pirate wouldn't kill me. The list went on and on. All I had to offer my son was fear and exhaustion. Ogden didn't deserve that." She paused. "Sandy was born to be a mother, and she was brave. If Micah was going to protect only one family out there, it had to be the Millers. They'd live as we had always wanted to. Purely."

Anika seemed to fall into thought, and Cal said, softly, "You don't have to tell us anything more." Frida wanted to smack him; he thought he was appeasing Anika, but Frida could tell that the woman wanted to unburden herself.

"You know," Anika said, "Peter went to see the Millers not too long before they died. I'd asked him to do it. Go check on our friends, I said. Please. I missed Ogden, and I suppose miss-

362

ing him made we wonder about Janie. I wanted to make sure she was okay. August said she was, but I never trusted him. And I guess I can't trust Peter now, either, because neither of them told me about Garrett. Their little happy family."

She laughed.

"What's funny?" Cal asked.

"You two are scared I'm going to shoot you when you could very well be poisoned in your beds. Give it three, four years."

"The Millers poisoned themselves," Cal said.

"Are you sure? Peter told me tonight that Micah went to see them before they died. Before that, Micah had never gone to visit the Millers—he preferred to act as if they didn't exist."

"I don't believe that," Frida said, though she did. Somehow, she knew Anika was telling the truth, but still she wanted to strangle her for speaking it aloud. There was only so much Frida could bear.

Anika cocked the gun. "Maybe I'm wrong. What do I know?" She looked at Cal. "How do I know you didn't kill the Millers?"

"I'm not a murderer," Cal said.

Anika raised an eyebrow. "How can I be sure?"

It was like Cal had been sucked into the air by some unnameable force, and Frida watched as his body swept the space between him and Anika. He grabbed the gun from her hands and slammed its handle against her face. Anika didn't make a noise as she fell to the floor.

"Cal!" Frida cried out.

He was already pulling off Anika's purse, yanking it so that the strap ripped free. "Take her flashlight, and wait for me at the bottom of the tree."

She dropped the book and took the bag. She was crying as she picked up the flashlight.

22

For once, his wife listened to him. When he heard Frida's feet hit the ground below, he turned back to Anika, who lay at his feet with a hand to her face. She was moaning softly.

"I told you to put down the gun," he said. He didn't know if she heard him. He still felt that clobbering rage. How dare she corner him and Frida and then suggest that he was the one to kill the Millers?

He was glad it was too dark to see her face clearly; if there was blood, he couldn't make it out. He didn't want to see what he'd done to her. She was moving a little, which meant she must be okay. He hadn't knocked her unconscious.

"I didn't kill anyone," he said.

She groaned, this time more loudly. With one hand, she pushed herself upright.

"Stay where you are," Cal said, but Anika wasn't listening. She was trying to stand up. He couldn't let her.

"Stop," he said, and raised the gun at her.

Anika kept moving until she was upright. Cal knew she'd never do what he asked, not now, not ever. And even if he and Frida got away, Anika might yell for the others on the Land to go after them. He needed to stop her.

Cal swung the gun through the air, and it landed against Anika's face for a second time. He thought he could hear the crunch of her cheek against the metal and, a second later, the crack of her head against the wooden floorboards. It was its own gunshot.

Cal felt his chest moving up and down, and he put his hand there, to still it. He waited to see if she would get up again, but she didn't. He knelt down and found her wrist and waited for her pulse. There it was; she was still alive. He hurried down the ladder.

Cal took the flashlight from Frida's hand and pulled her away from the tree. They couldn't run across the open field, or they'd be discovered. People might have heard the altercation in the woods, the noise traveling across the Land, shaking everyone awake. They would want to know what had happened. They'd be here soon.

If Cal and Frida went west, they'd eventually hit the Forms where they'd originally met Sailor. All they needed to do was get through those, and then they could return home. Maybe they'd spend a few hours there before heading out again. They should have that much time. Once the Land discovered Anika's injury, they might come after them.

"Please stop crying," he said as soon as they were on their way.

Frida was sniffling behind him, sucking in her whistling breath.

"It was either her or us, Frida."

He quickened his pace, pointing the flashlight into the trees beyond. Frida was right to be crying—what he'd done to Anika was brutal. He meant to slow her down to protect his wife and child, but in the moment he hadn't thought about what would happen to Anika after tonight.

"Please," Cal said, but he wasn't sure what he was asking for. Some forgiveness maybe, or just some space to focus. He had to focus.

From the Tower, he'd realized how small this stamp of woods was. They'd hit the Forms soon enough, as long as they kept walking in the right direction. He would not think of what had just happened, what he was capable of. He pointed the flashlight ahead of them and walked carefully forward.

The first Form had been built into the woods so that, from afar, it looked like another tree. It had even been built straight up, rather than curved, and it resembled a thick and sturdy trunk. When they got close enough to see what it was, Frida reached for Cal, and they stopped. The Form was filled with crushed rusty soda cans and other trash. It smelled vaguely of metal and sour milk.

"Not much of a Form," Cal said.

"You mean Spike," Frida whispered. She squeezed Anika's bag between her hands.

He nodded. "Spike."

They made their way carefully around the base of it, and as they did so, Cal said, "I saw a map of this place. Sailor showed it to me." Frida didn't reply, but she continued walking behind him, as if she trusted him to lead. She had stopped crying.

The second Spike came a few feet beyond where the woods stopped. Cal knew the broken-glass border wouldn't begin until they were a third of the way through, and he hoped by then he could get a handle on where they were. They just had to move forward—he was pretty certain they were moving west.

He didn't know what they would do once they'd escaped. Where could they go? He imagined Anika lying unconscious in the tree house.

"Just walk," Frida whispered.

The Spikes hulked over them. His flashlight shined on a bald bicycle seat, then a rocking chair with only one curved leg. Miniblinds, dented and bent.

He pictured them back in the Millers' house. It would feel empty without the things August had brought for them in the duffel bag. Frida would realize she no longer had her father's sweater or her abacus or her favorite dress. And then what? She would walk around the room with its metal table and crowded shelves, her eyes roaming. She would seize upon something new to treasure. They both would. Cal remembered his pillow, waiting on the bed in the Hotel, and something in him sagged. One more thing, lost.

It was cold and their breath was visible in the air. Cal had the gun in the pocket of his coat, and he felt its weight, and hated it.

They were walking fast, too fast, and Cal realized he didn't know where they were. Somehow they'd been on a descent, and now he couldn't see anything but what was right in front of them. A wall of Forms. Of Spikes. He didn't care what was in them, or what they were called. They didn't look familiar, they were built just inches apart, and suddenly Cal wasn't sure which way to go.

He grabbed Frida's hand. He was unable to admit that he'd probably taken a wrong turn. Even now he was a coward and a liar. "This way," he said, and turned them around.

They circled a smaller Spike, just taller than Frida, its tip capped with a saw, teeth sharp in the moonlight. He tried to see the maps in his mind, but there was only blankness. Maybe Micah had unwittingly told everyone the truth—he'd sent Cal and Frida out here to die.

"Frida," he said.

"I know," she said.

He knew she was just as scared as he was, and yet neither stopped walking. They had been lured into this labyrinth; they were under its spell.

Someone cleared his throat.

It was a man, Cal could tell by its gruff animal quality. Whoever it was, he was nearby, watching them, waiting. He would pounce.

Cal did not grab for his gun. Instead he said, "Run," and pulled Frida away from the sound.

They ran between the Spikes, turning one way and then another. The Spikes seemed to grow taller, and he thought he saw them swaying in a wind he didn't feel, bending to its will like trees and skyscrapers.

Frida was saying his name, but he didn't listen, he was dragging her as far as he could from that man.

They ran until Cal's arm caught on something sharp. Barbed wire. He felt like a piece of paper, torn in half. Frida cried out as if she, too, had felt the sting.

"Calvin." It was a man's voice.

He stopped, and looked behind them.

August stood with his arms crossed. He wore his sunglasses, even in this darkness. He didn't look cold. He looked like he had never been cold in his life.

"You'd better come with me," he said, and Cal knew it was over.

He gave up his gun and the flashlight, and Frida finally surrendered Anika's purse. In silence, August led them away, his own larger flashlight bobbing up and down with each step.

At first Cal thought they were headed back to the Land—for what, Cal didn't want to imagine—but when a few of the Spikes began to look familiar, he realized he and Frida had encountered them the day they'd arrived. So August was taking them away from this place.

Cal turned to Frida, whose face looked calm in the dark, even beatific, and he decided to follow her intuition. Cal wished he could feel as certain.

"Careful now," August said, when they approached the broken glass in the ground, but he would give nothing else away.

When they reached the last Spike, August said, "You're okay now."

"Thank you." Cal was ready to shake his hand, but August was gripping the flashlight and made no move to let go.

"Anika," Frida whispered.

"I know," August said.

"Shit," Cal said. "How did you find out? Does everyone know?"

August smiled. "You really think she went up there without telling anyone?"

"Why didn't you stop her?" Frida asked.

"Micah wanted it like that. He knew Cal would handle it. Anika was getting to be trouble for us."

"I only hit her with the gun," Cal said. "I didn't shoot."

"You did just enough," August said. He took one hand off the flashlight and patted Cal on the shoulder like he was a god-damned dog.

Cal's ears burned. "It wasn't a little trick I did for you."

"Please stop," Frida said, and Cal didn't know whom she was talking to.

"Can we go home then?" Frida asked. "Just like before?"

August shook his head. "You can't go back there, not with everyone on the Land so upset. It wouldn't be safe. Besides, the Millers' place is going to be occupied. Peter isn't happy with us. He thought we could convince the others to let in a child."

"And because Micah disagreed, he'd rather live on his own?"

August shrugged. "It's not Peter's choice to make. I suppose we'll find out what he'd rather do when he's apprised of the situation."

"I don't understand," Cal said, but he did. Micah would bend the Land back to his original vision. He had to get rid of detractors, Peter included.

"Where will we go?" Frida asked.

"The bus," August said.

"The school bus?" Frida suddenly looked scared.

"Don't worry," August said. "We'll talk there." He started walking again.

Cal knew immediately where they were going.

In a few minutes they reached the bus, parked in the middle of the field as if it were a perfectly normal place to find it. Cal wondered if the goldenrod they'd seen on the way in was still there. He'd find out tomorrow when the sun rose. Unless they were leaving right away.

August opened the accordion door. "After you," he said.

Frida climbed in first and, as if this were a field trip, sat in the second row. Cal followed her only when she tapped on the window and called his name.

"Hurry," August said.

The inside looked just as it should: the aisle carpeted with ribbed rubber, the rows of dark green seats, that close-body smell. Cal sat next to Frida, and she dropped her head on his shoulder.

August ascended and immediately went to the first row of seats. He pulled up the bench seat and from its innards retrieved a blanket and two cans of beans. "I have a can opener around here," he said. "After we eat, we'll get going."

Cal shook his head.

"You have to eat," August said. "You'll need the energy."

"Where are you taking us?" Frida asked.

"I'll tell you on the way," August said.

"Wait," Cal said. "I have more questions."

August sighed. "If it's about how we travel by bus, that's easy. I've got a vat of cooking grease in the back. This is a diesel. Again, I'll explain more once we're on our way."

"It's not that."

August raised an eyebrow, but he was listening.

"What happened to the Millers?"

"Oh, darling," Frida said. "Don't let Anika get to you."

"Stop, Frida," he said. "August? Tell me. Tell us."

August pulled out the can opener and handed it to Frida. She took it and squeezed the handles. She wanted the story, too, Cal saw.

"No one but me was supposed to have contact with Sandy and Bo," August said. "I'd told Micah about the boy, and we'd agreed to keep him to ourselves."

"Why?" Cal asked.

"Why not? Who would benefit from that information?"

"You wouldn't want everyone thinking they could just start a family out here," Frida said.

"When Sandy told me she was pregnant, I delivered a large canister of protein powder and a stethoscope and a kit to read her glucose levels. I had to risk a lot to get that to her. Without me, and the stuff I gave them—traded, my ass—there would have been no kid. Sandy would've hemorrhaged during birth. Even if they both survived that, the kid would most likely have died his first year."

Cal wondered if Micah knew what August had done for the Millers, or if keeping it from him was part of the risk.

"Peter went to see the Millers, without any of us knowing. Well, Anika knew. She requested he go."

"Was he angry when he found out about Garrett?" Cal asked.

"Like you wouldn't believe. He came back demanding to know why he'd been kept in the dark. He was in the morning meetings, after all, and Micah had always claimed to be totally transparent with him. He and Micah spoke privately, I don't know what was said. But it apparently hadn't totally appeased Peter because he came to me."

"In private?" Frida asked.

"He wanted to talk about the Millers coming back to the Land, with the two kids. It was absurd. Seeing Garrett must have flipped a switch in him—made him think about Ogden."

"What then?" Frida asked.

"Peter left the Land again, not long after we'd spoken."

August was now looking beyond them, as if waiting for someone to stop his story.

"He returned the next day. He wouldn't talk about it with me. He flung himself into things after that. Into morning meetings, governmental concerns. Even Micah's ideas."

"The plan," Cal said.

"What plan?" Frida asked.

"Nothing," August said.

"What is it?" Frida asked. She was looking at Cal.

"Just, you know," Cal said, his heart clanking in his chest. He couldn't tell her, she'd freak out. "Micah wants to build more Forms." It was a feeble lie, but it was all he had.

"Yeah," August said. He was talking to Frida, but his eyes were on Cal. "Peter got really into security issues."

Frida said nothing.

"Micah was angry when Peter returned the second time," August said. "They were in the tree house for over an hour, easy. Whatever happened on Peter's visit had Micah worked up. The next thing I knew, I was saddling up Sue, and Micah and I were going to Sandy and Bo's."

"Anika said my brother never went to see them," Frida said.

"She's right. He rarely spoke of them. He acted as though they were like any other settlers in the area. But when Peter came back from that second visit—"

"What changed?" Frida asked.

"Bo had seen something," August said. "He told Peter he didn't want him coming around anymore, that he was confusing Jane and Garrett. I guess it was Sandy who finally interrupted Bo and called

Peter a traitor."

"For what?" Cal asked. "What had Bo seen?"

"He'd taken to exploring by himself, way beyond the familiar territories," August said. "He'd seen a Pirate."

Frida shook her head. "But I thought the Pirates were gone, that you guys fixed that problem?"

"I wish it had been that easy," August said. "They still roam the wilder areas. They don't come anywhere near the Land or the settlers I'm in contact with, and they definitely stay away from Pines." He gripped the seatback that faced Cal and Frida. "They hold no real threat to us anymore, and there's a reason for that."

"So if they aren't a threat, why does it matter that Bo saw one?" Frida said.

"It matters because the Pirate wasn't alone," August said. "He was with your brother. Doing a trade, I'm guessing. Bo said they were smiling like old friends, though I'm sure that's Bo embellishing. Micah views his relationship with the Pirates as purely business—believe me, there isn't a kinship there. They're afraid of him, and they also depend on him, for supplies. A year or two ago, Micah killed one of their men. The kid was talking back, swaggering. Micah did it just to make sure they knew who was boss."

Cal waited for Frida to say something, but she was silent, staring at the mottled whirl of the seat upholstery in front of them.

"Let me guess," Cal said, "none of the original settlers on the Land know about this business relationship."

"Only Peter."

"Why does my brother trade with them?" Frida asked. "If he's not afraid to kill one, why not just get rid of them all and get rid of the problem?"

"The Pirates have their uses," August said. "They keep new settlers from coming in, for one, and they see stuff well beyond the territory I travel. Plus, if Micah were to totally eradicate them, he might be out of a job."

Cal didn't say anything. He remembered what Micah had said about the Pirates attacking Pines someday.

"So Bo found out about my brother's relationship with the Pirates," Frida said.

August nodded. "The Millers told Peter they were thinking of coming back to the Land to tell everyone. They thought people deserved to know. That had freaked Peter out. He'd be in trouble, too, seeing as he'd been aware of Micah's dealings all along. He told Bo and Sandy they better stay put, that Micah was working with the Pirates for everyone's benefit. The Millers were having none of it. After all the horror the Pirates had inflicted on the Land, what Micah was doing was unforgivable."

"And then Peter ran home and told my brother all this," Frida said.

"You have to understand," August said. "When we went to see the Millers, Micah's main goal was to make sure they never came to the Land. If they did, they'd topple everything we'd

374

worked for. He wanted them out of their house and out of our territory. He didn't see the benefit of protecting them anymore, and I didn't either. I'm not denying that. They were too much of a threat."

"What happened?" Frida asked.

August took off his sunglasses so that he could look right at Frida. His eyes were gentle for a moment. "They weren't cooperative."

"What does that mean?" Cal asked.

"It means they wouldn't leave, and Micah did what he had to do." August put his sunglasses back on. "Bo was strong, but not against two men, and not with his family watching. He didn't put up much of a fight because Sandy and the kids were there, and we hadn't touched them. Finally, he agreed to follow us outside." August paused. "Micah had brought the poison with him. I had no idea he had it. He wanted Bo to take it, and when Bo refused, there was a struggle. Micah held him down." He stopped speaking again, and Cal leaned forward. "To find signs of suffocation, you'd have to be looking for them."

"Jesus," Cal said.

"Afterward, Micah threw up right by the cart. He wiped his mouth and told me to wait with the mare while he went inside."

"Sandy didn't try to escape with the kids?" Cal asked.

"I thought she would have. I was praying she wouldn't. I didn't want to witness Micah chasing her, doing God knows what."

"You sat back and *watched?*" Frida said. "How could you?"

"I saw nothing," August replied coldly. "Sandy didn't scream or anything, either. It was quiet. She agreed to take the poison and give it to the children. She told them it was medicine. Later, Micah said she was just sitting on the bed with the kids, singing a song about birds, trying to keep them calm. It was all she could do, I guess, to get out of this mess." August paused.

"Why didn't he just let them leave?" Cal asked.

"Would she have wanted to go? She'd be out there in the woods, alone with her kids."

Frida wasn't crying, and the way her body felt next to his, Cal knew she was beyond crying, that sorrow had leached even that from her. Cal thought about his conversation with Bo about the Spikes. He must have been warning him away from this place.

When August started talking again, his voice was careful, measured, as if he'd reported this very procedure at the Land's next morning meeting.

"Frida . . . You should know that Micah wanted the Millers to vacate the house for another reason. He wanted you to live there."

Frida began trembling slightly. "Micah knew I was out here?" Her voice was higher than usual, almost squeaky. "Before I told you?" She was trying to catch her breath.

"As soon as I met you and reported your settlement to him, he knew."

"And he never came to see me?" Frida said.

Cal brought his arms around her.

"Micah wanted you out of the shed, and as quickly as possible," August said. "The shed was too small, he said, too hot in the summer and too cold in the winter. It was tiny. He knew you'd take the house." August paused. "He wanted you guys to have a better life."

Frida didn't answer.

"Micah didn't want me to bury the Millers' bodies," August continued. "He said that you"—here August looked at Cal—"would see the bodies and they'd scare you. That you'd never let

376

Frida leave. That you'd stick to routine as a way to survive. He said it's what you did in school, after your mom died."

Micah was right. He'd known that Cal would be a coward and that it would be Cal's job to hold back his wife.

Frida leaned into him. "It's okay," she whispered. Then to August, she said, "Why didn't you want us to think they'd been killed? That would have really scared us."

"That's what I said. But Micah was really freaked out, like I've never seen him. He didn't want you to be afraid, and he definitely didn't want you running away where he couldn't find you or heading here, looking for answers. He was panicked. I had to tell him to wait with Sue while I dragged Bo's body inside."

"You did it *for* him?" Frida asked bitterly.

Even though Cal understood why she was angry, he didn't want her to say anything more. She didn't get it; she couldn't. Micah might have committed the murders, but it had been August and Cal who had to go in afterward and face those deaths. Whatever had motivated August—loyalty or self-interest, desperation or fear, or even the same metallic coldness that ran through Micah—Cal couldn't blame him for following Micah, for tidying up the situation, for trying to resolve it in whatever way he could. Maybe what August had done was wrong, but it was human, too. Maybe it was compassion that had led August back into that house. He'd wanted to return Bo to his family.

No one said anything. Cal thought August looked tired, as if telling that story had been too much for him.

After a moment, August said, "Let's eat. It'll take us a while to get to Pines."

377

23

They'd left when it was still dark. Through the night and into the next day, they'd ridden over a road so rough that even Cal had thrown up twice into an old wastebasket August kept for such purposes. They kept stopping to move fallen trees out of the way. Once August had braked for a deer running across the road; at first he thought it was an Illegal. That's what they called them in Pines, he said. "Poor people, trying to get inside."

Frida wanted to ask, *Isn't that what we are?* but she didn't. It was another thing she'd keep to herself.

Frida had intentionally chosen to sit at the back of the bus; she knew it would make her sick, but she wanted to be as far as possible from Cal and August's conversation. They had spoken intently for the first half of their journey. Before starting the bus, August had told Cal he could still be of assistance to the Land. "Inside Pines," he said, "you'll have your eyes and ears on the ground." Frida could sense Cal deliberately not looking at her as August spoke. And could she blame him? Her brother had killed their friends, and August had stood by and let it happen, and Cal was still willing to work with them. But she understood. They were going to be parents. Cal just wanted to keep their family safe.

August had nodded at Cal and said, "There's a role for you inside. Micah's been thinking about it since you guys arrived."

Frida didn't know what they were talking about, but it was probably about Micah's plan. Whatever it was, it certainly wasn't about building more Forms.

Frida was relieved she couldn't hear them talking over the engine. She didn't want to know. She could only think about the baby and how they were being spirited away from Micah and the Land and the Millers and Anika's wounded head. They were headed for a safe place. It would be safe, right?

Cal was so invested in his conversation with August, he barely glanced out the window. Frida couldn't take her eyes off the world; it looked even worse than it had on their way out of L.A., as if, in their time away, it had slipped further from care.

The sky was gray above them, and the bus jostled over dirt and rocks and chunks of asphalt. Sometimes there were woods on either side of them, and sometimes the trees thinned out, turning black and shriveled, leading to soulless, empty spaces that reminded Frida of desolate parking lots. They once passed a gas station, dead. Someone had ripped out the hoses, and a fallen streetlight cut the convenience store in two. Occasionally, they passed houses, but they were falling over, deflated, covered in ivy, their doors knocked off, their roofs collapsed. A dog with xylophone ribs stalked an uneven porch. Frida hadn't seen a pet in years, but this wasn't anyone's pet. No one would return here. Not ever.

Frida kept thinking of Anika. How badly had Cal hurt her? And when Frida wasn't thinking about Anika, she was thinking about Jane and Garrett.

Whenever these dark thoughts came into her mind, Frida went back to the brochure August had given her. It made her feel better, boasting photographs of beautiful yet modest homes in a number of styles: California bungalow, Spanish Mediterranean, French countryside. They were so new, they seemed to glow off the page. *Thanks to our cutting-edge workout facilities and well-maintained bike paths, our valued citizens live active and healthy lifestyles. Just wait until you try our Good for You! Diet Plans™, offered in each of our six shopping districts.* The brochure said the population at Pines was capped at 10,500 for another five years or until the Community could build more neighborhoods and plazas to expand.

We want to maintain a small-town feel. Come home to us!

Curled into the back of the school bus, her legs smashed into the green vinyl seat in front of her, Frida had laughed at that. But she had to admit, other things excited her: *Enjoy a different cultural event every night of the week, from classical-music concerts to lectures on obscure typefaces.*

Cal would like that.

Take advantage of our speedy, hand-delivered correspondence system, and our Quality Interaction Centers™, where friends can meet face-to-face for stimulating conversation and a variety of antioxidant teas. At Pines, there is no time but quality time!

Pines residents hadn't eliminated technology from their Community; they just didn't celebrate it. *Once you're done with work, why spend time in front of a screen?* Maybe they lacked the resources, the satellites, the cell towers, but it didn't matter: they had transformed a flaw into an asset. Pines was supposed to remind you of a bygone world that no one living had seen first-hand: cookouts and block parties, paperboys and school recitals. Daddies who took the trolley home, mommies who put up their own wallpaper.

August had started the bus as soon as he'd eaten a little and filled the tank. As he turned on the engine, he'd said, "To everyone in Pines, Micah is dead. You *must* proceed from that notion."

Cal had looked at Frida then, as if she might not want to

cooperate, and she'd nodded. Giving up her brother once more was a relief, actually.

They had seen no one else on the way. "God willing," August said, "it'll be just us until we reach the first checkpoint."

As they approached Pines, August admitted to getting permits for Cal and Frida on his last trip. He hadn't acquired entrance papers in years, he said, and Frida thought he must be thinking about the children he'd transported from the Land. "I got them from Toni," he said.

"Toni?" Frida said.

"She's on the inside here," Cal said slowly. "I should have told you."

"Too late for that." Frida was supposed to be upset, and she was, but she'd learned so much in the last few hours, this couldn't compare. Toni. Frida couldn't suppress a smile. Her old friend.

Now Frida and Cal were a young couple, recruited to come to Pines because Frida—now Julie—was with child, and because Cal—now Gray—could be of assistance in their Education Department. To Cal, August had said, "You'll be able to get to know those inside who are assisting us. And you'll see who might be questioned by authorities, should trouble arise."

Frida shut her ears.

The gates to get into Pines were tall and ornate. Like the gates of heaven, Frida thought stupidly. The man who pored over their paperwork was wearing a white button-down shirt so clean it made her eyes hurt.

"Antonia Marles preapproved them," August said to the man.

Either Toni had pull, or there was a lot of gold in the small cashier's box August handed the inspector, because they were let in after just a thirty-six-minute wait. Frida knew because she'd kept her eyes on the clock tower a few feet ahead. She'd been so nervous, she thought she might barf into the wastebasket. But they'd been waved through, and August parked in the nearest lot. Beyond that, he told them, were only the occasional delivery trucks and electric station wagons, which the richest families drove. Everyone else walked, took the trolley, or rode bicycles.

"It's nice," Frida had said as they turned onto their street. She meant it. Beyond the border, there were at least three miles before the world started to look wild and ruined and frightening. Here everything had a start-over feeling.

24

Frida was still in bed, hoping she would fall back asleep. She wanted to spend the morning here and wake only for lunch, well rested. That plan never worked—now that she was eight months along, her belly large and ungainly, she could never stay comfortable enough to sleep the day away. But maybe it'll work, she thought, just this once.

The baby kicked twice, as if to say, *Ha*. Frida rolled onto her other side and pulled the duvet to her chin.

Cal was in the shower; he had five minutes before the timer went off and the hot water turned cold again. She could hear him singing the song about the doggy in the window; it had been playing the last time they went to the Central Shopping Plaza. His voice sounded decent—deep yet sweet—but it was so unlike him that she laughed. How appropriate. He wasn't Cal anymore, anyway. She was supposed to call him Gray.

Frida had not been sleeping well since they'd arrived. Apparently, a memory-foam mattress and two semifirm hypoallergenic pillows couldn't help Frida sleep through the humidifier sighing on its pedestal across the room or the bothersome glow of streetlights outside. The fridge downstairs liked to make ice at two in the morning, and at dawn, the paperboy came down the block on his bicycle; Frida could hear him every day as he pulled copies of the newsletter out of his canvas bag and threw them onto the lawns. From the bedroom window it looked like a tiny white flute had been tossed onto the dewy grass below.

Last night had been especially tough. The police had taken the neighbor's son away. They'd been stealthy about it, or as stealthy as you can be when you're throwing a grown man into the back of an armored vehicle. Mrs. Doyle's son had been picked up because he was a drug addict. That was obvious; his downward spiral had saddened the whole neighborhood. For five years, Frida was told, he'd been the basketball coach for Pines West. Two years ago they'd won the championship, played against five other Communities in the Western and Northwest Territories. Frida had gotten used to saying that: *the Territories.* They were technically still part of the U.S., but those two letters sounded so antiquated, so inaccurate.

"It's not like the United States government contributes anything to our well-being," Cal once said to Mrs. Doyle. He was very good at repeating rhetoric from the newsletter, but in a way that sounded fresh. "They can't tax us if we don't depend on them for anything. Besides, the government's a wreck."

It had been clear that Mrs. Doyle's son wasn't doing well. For one, he kept skipping practice—his players dribbling balls listlessly for an hour before heading home—and sometimes he'd go missing for days. Some claimed that the latter wasn't true, that Mrs. Doyle had been hiding him in her basement, to protect him from his urges, or to keep anyone from seeing him high. But that couldn't be, because she was the one who had told the authorities. "It's best for him," she told Frida. "And for everyone. We can't have drugs crossing into our borders."

Frida didn't know if it was heroin or crack or what, but people were getting it into Pines, and the authorities wanted to eradicate the problem as quickly as they could. Their philosophy was: get rid of the customer base to get rid of the product. Anyone who didn't follow the rules was thrown out. Easy as pie, people said. The good residents of Pines weren't paying luxury rates to live among criminals. The damaged ones were not welcome.

She ran her legs against the sheets and plowed her cheek into her pillow—she couldn't get enough of this pillow.

Someone had to be doing undercover work to try to identify the smugglers, Frida thought.

That, or the authorities were funneling the drugs into Pines themselves. They would sniff out the weak ones, the Problem Children, as everyone called them, no matter their age. She wanted to ask Cal about it, but he was worried—paranoid maybe—that the cameras and microphones on every street corner would pick up their conversations, even inside their home, and so she kept quiet. If she wanted to talk to him about anything serious, Cal would only allow it if they sat on the fuzzy blue mat in their downstairs bathroom with the shower running. Besides the walk-in closet, it was the only room in their house without a window, and Cal preferred it because they could run the loud ceiling vent to further mask their conversation. Frida had to lean forward to hear what Cal was telling her.

The signs were posted at every corner: smile, you're on camera. She used to see that same message at liquor stores and trashy clothing shops in L.A., except these had been custom-made for Pines, the words written in a navy-blue cursive; beneath them a doll stared out with large, blue eyes. If Frida didn't know better, she might want one of those signs to hang above her fireplace. They were stylish. The Design Department at Pines was the best in the Western Territory.

The first time Frida saw a curb your dog sign, it made her weak with glee: black dachshund, green leash, the letters in that same bubbly cursive, but this time fire-engine red instead of navy. It had been their first week as newly minted citizens of Pines, and they'd been walking through the park nearby, holding hands. They were taking it all in: the cameras, the signs, the park's plastic castle with its slide and shaky rope bridge, the kids squealing as they crossed to safety on the other side.

Frida was in her final trimester now. Another reason why she couldn't sleep. She had to lie on her side all night long when she used to sleep on her back, and sometimes it felt like the baby was doing acrobatics off her ribs. The doctor had offered to tell them if it was a boy or a girl, but Frida had refused. Every time the baby's image filled the sonogram monitor, she looked away. It didn't matter what the machine showed, or even that Cal hoped they were having a boy. Frida knew it was a girl.

"As long as it's healthy," she said, which was what every pregnant woman was supposed to say. The doctor had nodded, said of course the baby was healthy, and reminded her to continue taking her vitamins, her daily Protein Shakes, her weekly SuperFoods Package. "Don't skimp now," he said kindly.

Cal complained openly about the food here. A lot of people did. Cal didn't seem worried if someone recorded his displeasure with the Vegetable and Fruit Pills, and the Shakes that tasted like chalk even if their containers read CHICKEN PROTEIN OR BEEF WELLINGTON PROTEIN. He worked in the Education Department, after all, and one of their ongoing concerns was school lunches and the nutritional needs of children. Cal seemed comfortable taking issue with the supermarkets here, which had aisles upon aisles of powdered foodstuffs and the smallest, most pathetic produce sections. Greens were expensive, and there wasn't much else to choose from beyond carrots and the occasional tomato. On an early visit to the market, Frida had run a piece of lettuce between her fingers, imagining that it had been grown on the Land. But she'd only done it once. She had better restraint than that now.

Cal was going to start a garden soon, as others had done. There was very little room in Pines for a massive agricultural movement, but Cal said there was talk of redoing one of the shopping plazas to include a small farm. Apparently, when Pines was first being built, its developers had overlooked the value of fresh produce to its clientele, and now they were rethinking the matter.

There was no real meat anymore, either. "Mad cow, E. coli," Mrs. Doyle had said with a frown when Frida had asked her why not. "Plus, the amount of water required to raise a head of cattle? Forget about it!" Cal said he'd gladly be a vegetarian if he could eat some vegetables. Frida was just relieved the baking section at their nearest market was so stellar. Pines wanted every wife to be a cake baker, wearing an apron and a smile.

"Maybe I can become a pastry chef," Frida had said to Cal once.

"I don't know about that," he'd replied. "This place is specifically designed for a certain kind of family. You know, the father at the office for long hours, the mother busy with the kids."

"What do those fathers do in those offices all day anyway?" Frida had asked. It was still a mystery to her, how Pines worked. Meanwhile, all the mothers stayed home to bake cakes and whatever else mothers at Pines did. Women were expected to devote everything to raising a family.

"You know, taking the older kids to art class and soccer practice," Cal said. He was just reciting the brochures from memory, but his message was clear: *We're going to be parents soon. Let's blend in.*

Their baby would be born in Pines Hospital, and then she'd bring her here, to this beautiful house, and place her in this clean bassinet. She'd tell her stories about her family in Los Angeles, a city that was too dangerous to visit. She'd tell her how Mommy and Daddy had received permits to come to Pines. *Daddy got a job because he's so smart,* she would say. *And they wanted a baby like you to live here.* It was an easy myth, and she practiced it in her head every day.

The bassinet was already here, next to the bed. She had painted the nursery a pale yellow, but it remained empty. She hoped the furniture store in the Central Shopping Plaza might send a messenger today, to tell her they had finished building the crib.

From the bed, Frida heard the water stop, and then Cal pull his towel off the rack. Their closet was attached to the bathroom, as big as a room, and she imagined him facing his row of clothing, sliding the hangers across the rod. They had both imagined well-cut suits for him, handmade by the local tailor, the fabric at once sumptuous and crisp. But no. The suits he received were cheap: synthetic fabric that wrinkled easily and didn't breathe well. Frida's dresses were just as bad, and after the first washing they began to pill and fade.

What they got had been pricey, Toni informed them later.

She had them meet her at their local center, for tea and stimulating conversation. Even Toni had used the brochure language.

"Not everything can be luxurious," she'd told them when Frida made a joke about Cal's khakis. The hems were uneven. People said the same about the lack of hot water and the days when their electricity shut off completely. Residents knew ahead of time when the lights wouldn't work, but still, it was shocking to sit in the dark in their big new house. It made Frida afraid, just a little. Pines wasn't perfect; things were running out here, too. Mostly, she didn't worry, though; almost every day, the newsletter mentioned advances in alternative energy. They'd find a way.

That day at the center, Toni had stirred honey into her tea. They sat by the window, which overlooked yet another park, this one studded with benches and large outdoor sculptures. A man and his son were flying a kite that looked like a strawberry.

"In Phoenix," Toni said, "they've got a Community that has a haberdashery and a man to make your shoes. But everyone lives in these terrible apartments, clearly designed by a cheap-skate."

She smiled and lifted her mug; it was one of those ceramic ones without handles, and Toni held it gingerly. It seemed the only thing she approached with hesitation.

It had been a shock to see her. She wasn't same person Frida had known years ago. Toni's hair was cut to chin length and dyed light blond, and she sounded different, too: her voice bright and cheery. People kept approaching her, to say hello, to say they wanted a meeting with her, and she kept passing out her business card, nodding to her assistant Gregory, who sat across the room, writing in a notepad. She was in charge of Citizen Compliance and Outreach. Frida had said it sounded like a big job, but Toni had only shrugged, fake modestly.

"Do you still go running?" Frida had asked her as they stood up to leave.

For a moment, Toni looked like she'd been slapped. And then she smiled. "Oh dear, not since middle school, I don't think." She held her gaze on Frida, as if to say, *Watch it,* and Frida had nodded, trying her best to smile. August had told her to pretend they'd never met Toni. "Ms. Marles," he'd called her, right before they approached Pines. Frida should have listened; it was naïve to think she could get her friend back.

Now it was only Cal who saw Toni, at work Downtown. He

was advising on a number of matters, including the nutrition thing, but mostly he was helping to design courses for Pines University. The school would soon open its doors to citizens from neighboring Communities, should there be spots open. The Communities were interacting more and more every year. It was the only way to maintain safety, and their trade programs made them more alluring to prospective clients.

Frida hadn't asked Cal anything else about his job, or about Toni's, though she was curious how Toni was allowed to hold such a prestigious position in a Community so obsessed with the idea of women being homemakers. Cal had told her that Toni was married to her work, but Frida wondered if the real reason Toni was single was because she still held a torch for Micah. Or maybe she and Micah were still an item. Possibly, but Frida didn't bring it up with Cal. If anyone at Pines were to ask her for information, she wouldn't have any; if she were truly ignorant, she'd be better off.

Cal was still singing from the closet. It made her think of Rachel, belting it out at the campfire. They were probably all happy that the winter had turned out so mild, and that spring was just around the bend. She wondered how Anika was doing. She wanted to believe the place had returned to normal, that she and Cal had been forgotten.

She pictured Peter cooking on the outdoor stove at the Millers' place. And then she imagined her brother, standing on the Church stage. He was alive, but she had to pretend he wasn't. She wanted to pretend that. It was the price she paid for this large, sunny room with its fluffy cream-colored carpeting, its big bed, with side tables to match. Or no: the price was the mild but nagging guilt she felt every morning, waking to this new life.

Frida sat up. If Cal was going to emerge from the walk-in

closet fully dressed, maybe even with his shoes on, she should at least drink some water and chew a piece of Refresh! so that she didn't have morning breath. She swung her legs to the side of the bed and placed a hand on her stomach, which was so big and round it looked silly, like she'd swallowed a globe. The veins beneath her skin were like river lines on a map. Her belly button had popped weeks ago. The baby danced inside of her.

She thought of Ogden. Every time she passed a herd of schoolchildren, she imagined him among them. Sometimes she even pictured him living on this street, riding the trolley with his adoptive mother, getting lunches at school. He was doing well; he had to be.

When Frida saw someone in the telltale navy-blue uniform, spraying vomit off the sidewalk or cleaning the public bathrooms, she thought of the older kids who had been brought from the Land. They were Hatters now, and one day they'd be wearing these uniforms, doing these jobs.

Frida had seen the Center at Pines just once. It was a big beige building on the south side of the Community, surrounded by well-manicured trees and a sloping lawn too radiant to be real, as unscathed and bereft of human commotion as a corporate office park. The newsletter sometimes did stories about the Center, about the services the Hatters were learning to keep Pines clean and running smoothly, and Frida kept waiting to see a group of them training with older workers. She hadn't yet.

She hoped the kids from Pines were okay at the Center, happy even.

Again, Frida thought of Mrs. Doyle's son. What was he eating now? His had to be a terrible exile, wherever it was. She pictured that skinny dog they'd passed on the ride here. She and Cal, they were lucky. Frida knew she was thinking only of her own family, that she had begun to see them as special: separate from the rest of the world with all its attendant suffering and corruption. Maybe it was wrong, but it was the choice she had made.

He emerged from the closet humming. He had the blue suit on, the least egregious of the bunch, and a red-and-blue-striped tie. The collar of his white button-down shirt was stiff, and too pointy, but even so, he looked handsome. His hair was cut short now.

They didn't discuss his long hours. Or the talking-with-the-shower-running. Or how tightly he held her at night, arms straining to reach across her belly. On some days, there was a furtiveness to his movements, the way he looked left, then right, as he approached the house, the way, when one of Toni's messengers came with new Correspondence, he let out an unnatural guffaw. "Well, here you are!" he'd say, like someone's pathetic uncle. On other days, he acted so smooth and comfortable here that the world settled around them like water, filling the empty space. On the smooth days, it was as if they'd always lived this life.

If at other times, things felt a little off, so be it. If something seemed wrong, if it seemed like he had something up his sleeve, she could ignore it. Whatever her husband had agreed to do, he had the best interests of their family in mind.

Her job was not to ask any questions. She and the child, they would stay here.

"Good morning, beautiful," he said now, coming toward her.

Julie stood up to kiss her husband. "Good morning, Gray," she said.

ACKNOWLEDGMENTS

I am indebted to the amazing Allie Sommer for her sharp and wise editorial guidance, and to everyone at Little, Brown for their support and enthusiasm. Thank you as well to Clare Smith and the whole team at Little, Brown UK.

My agent, Erin Hosier, is the queen of all agents. Thanks, E., for your honesty and humor, and for believing in me and my work way back when.

I would like to thank the Ucross Foundation, where I started this novel, and Hector Arias and Karla Escalante, who let me write portions of this book in their apartment while mine was taken over by a new baby. Thank you, too, to everyone who provided child care during that time. Your help meant (and means) the world to me.

Eternal gratitude to my wondrous and loving family: my father, Bob Lepucki; my mother, Margaret Guzik; my stepfather, Mitchell Guzik; my sisters, Lauren Lepucki Tatzko, Heidi Cascardo, and Sarah Guzik; my brother, Asher Guzik, whose assistance with all things plant related was enormously helpful as I revised this novel; my "stepmother" Keitha Lowrance; and my in-laws, the Browns. I am so lucky to have you all in my life.

I am grateful to the following people for their insightful feedback on this book: Madeline McDonnell, Mike Reynolds, Julia Whicker, Kristen Daniels, and Cecil Castellucci. Thanks also to

Dan Chaon, Emma Straub, Rachel Fershleiser, Ben Fountain, and Shya Scanlon for reading the complete draft and offering their support.

Thank you to Deena Drewis at Nouvella and C. Max Magee at *The Millions*. Thank you to Oberlin College, the Iowa Writers' Workshop, Book Soup, and Skylight Bookstore.

Thank you to Kiki Petrosino, whose poem "Valentine" inspired the Official Pussy Inspector T-shirt.

Thanks to my colleagues and students at Writing Workshops Los Angeles. I am especially grateful to my students who have shared their novels with me. I've valued our conversations immensely.

For their friendship, I'd also like to thank Diana Samardzic, Douglas Diesenhaus, Molly McDonald, Christine Frerichs, Kathleen Potthoff, Allison Hill, Stephanie Ford, Charlie White, Joshua Yocum, Laura Shields, Kirsten Reach, Paria Kooklan, Ryan Miller, Anna Solomon, Lisa Srisuro, and Michael Fusco.

Lastly, thank you to my boy, Dixon Bean Brown, and to my man, Patrick Brown.

Patty, thank you again and again. And again.